Working With
Hard-to-Reach Youth

Working With Hard-to-Reach Youth

Jillian Burkett

ISBN 978-0-557-71152-9

To protect confidentiality school locations, client names and identifying factors such as race, gender, age and family information have been altered. In some instances, students have been combined to create a "typical" case. Only key psychological symptoms necessary to the case at hand were maintained to illustrate the topic of discussion.

NOTE: The publisher/author of this work has made every effort to ensure that the drug procedures are accurate and in accord with standards accepted at the time of publication. As medical research and practice advance, however, therapeutic standards may change. For this reason it is recommended that readers follow the advice of the physician who is involved in the direct care of the patient/client.

Cover design: Lulu Publishing
Cover Photographs: Patricia Marks (left), Margo Petrowski (right)/Bigstock.com
Author Photograph: Erica Freed

Acknowledgements

This book could not have been made without the input, feedback, and support of John Moreno, Rosaleen Fitzpatrick, Pasty Clay, and Jane Shaffer, who each in their own way encouraged me, providing their own unique perspective on my manuscript. A special thanks to Jennifer Colvin, whose encouragement and late night phone calls have kept me going through the darker times.

Additionally, this book would not have been published without the undying belief and help of Ray Martino. His work and devotion to special needs children has no bounds, and I am but a beneficiary of his generosity.

Finally, this book would never have been possible without the experience I have gained through the work with my clients—for that, I must thank each and every client whom I have worked with over the past decade for sharing their lives, secrets, and trust with me. Without you, I would not be the therapist I am today. My clients have been my teachers, and it is because of them I am truly wiser and emotionally richer.

This book is dedicated to every child who took a risk and shared their life story with a therapist, or trusted a teacher and asked for help.

Table of Contents

Introduction

When thinking of a title for this book, I rejected many terms. At-risk youth was too general. In today's society, every child is at risk for some type of psychiatric problem or another. When I was growing up in the 1960s, I was able to stay outside playing after dark and my parents didn't have to worry. In fact, over 70 percent of my play was probably outside. I climbed trees, roller-skated, hoola-hooped, and pogo-sticked my way around the neighborhood. I rode my bike and walked everywhere—to school, my friends, even the nearby mall when I was older.

Now, even middle-class children whose parents are still married are at risk of obesity due to cable television, internet, playstations, and a multitude of other electronic devices that at the time of this publication will have been upgraded twice in order to entice the kids to buy the newest version. In addition, the current economics usually require both parents to work, often resulting in meals consisting of fast-food, or other less healthy options. These kids are at risk for obesity, which causes teasing in school, resulting in low self-esteem and depression. This can then result in children staying home from school to avoid the teasing.

The news is filled with internet predators preying on children, and teens are getting jail time for sending porn over cell phones. Children are committing suicide due to online bullying. So who is not at risk?

Similarly, the words "challenging" or "difficult" were too negative. The reality is that these words reflect some degree of responsibility on the part of the clinician. Imagine a room full of students who are given a complex math problem. Half the students finish the problem quickly and say it was a "snap." The other half struggle with the problem, giving up and saying it was "too difficult." Is it the math problem? Or the student?

Every clinician will find their niche—that population which excites them and motivates them. For these clinicians, that population would never be described as "difficult." For me, there are certain populations I would not touch with a ten-foot pole. Eating disorders, for example. I worked with enough female clients in private practice who had borderline eating disorder tendencies to know that an eating disorder facility would make me run for the hills. But give me a teenage gang-banger from a broken home, and I'm in my element.

So why did I chose the phrase "hard-to-reach"? I could have used the word mandated; but that sounds technical and legal. Many of today's youth who are in a group home, non-public school, or residential facility have an entourage of therapists. They have a social worker, psychiatrist, outside therapist, family therapist, school therapist, and group therapist.

Imagine you work in a corporate office and have to attend one of those cookie-cutter seminars on sexual harassment or time management. You're there for eight hours, bored. Maybe you listen, maybe you don't. You make a grocery list, a to-do list, you converse with your neighbor, or text a friend. Now imagine you have to go to these seminars every week for an entire year. Would you ever listen? Or would you just tune the speaker out?

Working with these kids is often like that. They have been counseled to death. They tune you out. You become the adult in a Charlie Brown cartoon, and all they hear is "wa-wa-wa," when you talk. I often joke with my kids that I could be talking to the wall, because they just aren't getting it.

Private practice patients come to a therapist because they want your help. They are paying for your services. These kids come because they have to. They didn't get to choose you, and they get punished if they do not participate. Are they wrong to resent you? Not all of the kids I've worked with are like this. Most of them love coming, and once they figure out who I am and how I work, we have a great relationship. Even the ones who claim, "I don't need no fuckin' therapist!" when first introduced to me can usually be turned around with the right persuasion.

So for whatever reason you bought this book—because you already work with this population, or are considering working with these kids—all I can say is: Hang on to your hat, it's going to be a bumpy ride.

1

Why Bother?

Andy, a young black man at my school in the San Fernando Valley, came onto my caseload when he was ten, and we worked together for over four years. For the last two years in his IEP (Individual Education Plan) his father would bring up the possibility of Andy returning to public school, and I would give a firm no, stating that Andy had not passed his counseling goal. Andy was academically bright, and could easily have handled the transition to public school on the academic side, but his weakest area was his social skills. At a larger school, Andy would be a clear target for bullying, and I felt he would fall into the "danger to self" category. For two years in a row Andy cried as I explained he would not be returning to public school.

"I can't take it anymore!" he would scream, "I hate this school. It's for retards."

When Andy reached the ninth grade, I saw him start to finally mature. He was in the high school building with eleventh and twelfth graders as role models, and Andy began to participate in counseling in a way I hadn't seen in the past. He began to develop insight into his own behaviors, was able to solve problems, and see how he could act differently without having to rely on the staff every five minutes to reassure him or guide him.

After the winter break, I went to the school psychologist, and informed her that Andy's IEP would be coming up in the spring, and I wanted to make it an Exit IEP. I felt that Andy was ready to return to his home school. He had been in the NPS (Non-Public School) system since he was five. I talked to the necessary administrative staff who notified Andy's parents. Meanwhile, we did not tell Andy. The kids in our school have a vindictive and jealous nature—when they learn someone is leaving, similar to in prison, they try to sabotage the student's chances by instigating them, hoping to get the exiting student to "act out" before they leave so that their home school will change their mind and not take them back.

However, as Andy's IEP approached, he became more and more agitated, and I decided I needed to tell him what was happening, before he sabotaged himself through his own insecurities. About one month prior to his IEP, I went to his class and called him and his teacher into the hallway. Andy looked scared—most students who get called out with a teacher are in trouble, and Andy normally had a tendency to see the glass as half-empty.

"You're not in trouble," I reassured him. "I just wanted to let you know, that back in January, I spoke to the staff, and we're in the process of exiting you. At your IEP, your home school district will be present. We're planning on having you return to your home school for summer term, when it won't be a full course load."

If it had been physically possible for Andy to jump through the ceiling, he would have. He slammed his clenched fists into the wall, and thumped them several times. His teacher patted him on the back, telling him to calm down.

"No, no," Andy said, "I'm calm. Do you realize, this is the best fucking day of my life?"

Andy's teacher and I spoke to him about keeping this information to himself so his peers wouldn't sabotage him. I reiterated that he was not to even call former students who had already left the school, because they could call up students who were still here. I joked with him, "So wipe that smile off your face, because you look suspicious as hell."

Still agitated, Andy asked if we could go somewhere quiet, where he could be alone. We went to my office, and Andy burst into tears. Crying, he just kept repeating that this was the best thing that had happened to him ever. This was what he had waited years for.

"Well, you earned it," I said. "I told you last year that if you proved you could handle yourself with the high-school students, you would go back. You should be proud of yourself."

Of course, as the reality of what was happening sank in, Andy would spend the rest of his time with me talking about his fears related to leaving the environment he had known for most of his academic life. He had not been in a public school since kindergarten. He was going from a school that housed approximately 400 students and ran grades 1–12, to a high school with over 4,000 students. He was going to have to deal with changing clothes and showering during physical education. He was going to have to catch up academically; his As at an NPS school would equal Cs in a public school. But I was comfortable in the knowledge that Andy now had the skills to handle these stressors without behaving the way he had when I first started working with him, which was to scream, whine, and break things.

I have had many children that I worked with return to public school. If they do leave, it is usually after I work with them for two years. Some come directly from public school, some come from another NPS school. But I was never as proud of anyone as I was of Andy. Andy had the deck stacked against him, because he had been at our school the longest, and had no outside services—no family therapy, no social worker, no one to assure his family complied with treatment. Andy did the work himself. Alone.

On the flip side of the coin was Ahmed. I worked with Ahmed for three years. When he first came to my attention, he had been misdiagnosed autistic, and was about seven or eight. I corrected the diagnosis, and he was put in a more appropriate class. Ahmed had great social skills, but was unable to read—not even two-letter words. However, Ahmed was great at play therapy, and would create elaborate scenarios using legos and talk aloud creating stories in session. He also loved to draw and do art work, and did things creatively with his pictures that other children never thought to do. For example, he would draw a house, then ask for scissors to cut the doors and windows so that they actually opened in his drawing, creating a 3-dimensional effect.

When Ahmed was first enrolled, he was living in foster care. He had a social worker, but his parents retained educational rights and came to Ahmed's IEPs. They did not speak English, and there was no translator on staff who spoke Persian which made communicating with them difficult. After approximately 18 months Ahmed was returned to his family, and that's when things began to change.

Ahmed started swearing in session, his attitude became defiant, and he started stealing regularly. He stole toys from my office, and supplies from the classroom. He admitted to stealing from stores. He shut down emotionally. He came in regularly with bruises from "falling out of the bed." Many child abuse reports were filed, but nothing seemed to change. Then Ahmed didn't come to school for weeks, and the attendance record showed he was hospitalized. When he returned, he was sedated to the point of somnolence. He was sleeping at his desk and drooling. He was unable to walk up the stairs to my office without staggering, walking into walls, and slurring his words.

I filed a child abuse report, and the following day Ahmed came to school coherent. Whether the parents were deliberately giving him too much medication to control his behavior, were unable to read the prescription bottle which was in English, or were giving him the right medication and just did not care that it had obvious harmful side effects I never found out.

Months later, Ahmed came to school, and the details of what happened were unknown to me. But he was immediately transported to a hospital after arriving on the school bus and I suspect the parents again had somehow abused him—intentionally or otherwise. When he came back from the hospital, he again had bruises, and continued to claim he "fell out of bed." This time claiming he fell out of the hospital bed, which is of course impossible since they have side rails.

Approximately one month later, Ahmed came to school covered from his neck to his abdomen in bruises which were approximately three to four inches in diameter each. I found a camera to record the bruising, and contacted the police directly. It turns out, Ahmed's other therapist had already filed a child abuse report for the bruising, and I got an annoyed phone call from the Los Angeles Police Department.

The officer who had gone on the call to Ahmed's home wanted to know why I filed a report when one had already been made. I told him that was the law. The officer told me they did not remove the child from the home, because he was ten, and did not say he was being abused.

"Are you kidding me?" I shouted into the phone. "Have you seen his bruising? That's abuse! Those are restraint marks!"

"He says he fell out of bed," the officer said in an exasperated tone.

"Well, he told me he walked into a wall—changing stories is a sign of abuse. The only way he could have gotten those bruises from falling out of a bed is if it was a bunk bed and he bounced—fifty times! Did you even look in his bedroom? If he is falling out of bed that often, his parents are responsible for neglect by not addressing his injuries and getting a different bed."

The cop could not be bothered. He believed the parents' story, and even though I had worked on the case for three years, nothing I said—and I talked to him for thirty minutes—could persuade him that this child was in danger. My last words to him were, "Fine, then on your head be it when this child ends up in an ER with a broken spine, or worse, ends up dead."

Less than four months later I arrived at work on a Tuesday morning after a three-day weekend, and a colleague met me in the hallway with a grieved look on her face. "Ahmed M- is dead. He died on Sunday."

I often joke with people that I have a bi-polar job. The highs are very, very high; but the lows are very, very low. I think these two cases illustrate that perfectly. So why bother going into this business? For the record, I should state that I only work at the school approximately twenty hours a week, and I have another job doing something entirely different where I work with adults. This is because the school I work at does not pay when school is out of session, so clearly I need some additional income when school is closed. However, I have worked at a school full-time, while also working at a group home. When I was an intern, I worked at a school in Watts full time. I have also worked as a supervisor of one-to-one aides of special needs students within the public school system.

So back to the original question—why choose this career? For one, it can be rewarding when done well. For myself, I like the unpredictable quality. It is similar to working in a hospital; you never know what is going to pull up in the ambulance. A teacher may call me and say a student needs to talk to me and I may have to do crisis work. A student may need to be hospitalized. They may have a meltdown in class and need talking to. Or the day could be "normal" with kids just acting out the way they regularly do. A student who was well behaved in session the prior week may act out the next week, or refuse to come at all for no apparent reason. It's sort of a crap shoot. If you like a job that's orderly, and you need to have a sense of control, this is not the right population to work with. It requires flexibility, and an ability to go with the flow.

According to the 25th Annual Report to Congress on the Implementation of the Individuals with Disabilities Education Act (2003), as of December 1, 2001, there were over 5.8 million students between the ages of 6 through 21 being served under IDEA. Between 1999–2000, over half of these students received family training, counseling or other support (56%) and/or psychological services (51%). Nearly half received social work services (49%) or one-to-one paraeducator/assistant services (45%). According to the AFCARS Report, 510,000 children were in foster care in September 30, 2006. In 2005, 556,500 children were placed on probation, while 140,100 were placed in residential or other restricted environments due to their delinquent behaviors where counseling services would be presumably be provided (Juvenile Court Statistics Databook). Obviously these numbers overlap—kids in special education settings can also be in foster care, and kids on probation can be in foster care *and* special education schools.

But the staggering numbers speak to the need for special education teachers, special education aides, one-to-one aides, child care workers for residential facilities and group homes, probation workers, social workers, and therapists. If you choose to work with adolescents, especially the at-risk group—you will never be without a job. Working with certain demographics has limitations during times of financial recession or instability—but emotionally disturbed children are not going anywhere, and their numbers appear to be growing.

In a non-public setting, as soon as you lose a client, another client is there to take his or her place. When I worked at a group home, kids came and went. Sometimes they only lasted a week before they AWOLed and were picked up by the police. A week later, a new face would arrive. I never have to worry about where my next client will come from, a concern people in private practice have to consider when a client leaves.

I do not have to worry week-to-week if my clients will show up—where are they going to go? In a school setting, unless the child has truancy issues, you can expect them to be there. In a residential or group setting, again—they are going to be there. Many of my private practice friends complain during the Christmas or summer months because their patients do not have the money to come due to the holidays, or they go on vacation. Well, my kids do not take vacations, and if they do, it is when school is closed and I'm able to increase my work at my other job to compensate.

More importantly, since when a child leaves another child is immediately available to fill the spot, I do not have to worry about hanging onto a client long after they have stopped needing therapeutic services. So many private practice therapists I have encountered, especially those from the 1970s, seem to keep their clients in therapy for twenty years on average. My feeling is, unless your mother was Lizzie Borden and your father was Hannibal Lector, who needs *that* much therapy? Working in a school, with an endless supply of clients, I do not have to worry about crossing that ethical boundary, because I know for every child that I recommend is ready to leave the school, there will be another to take their place. I can honestly work in the best interest in the client—not my pocketbook.

Every therapist has their own level of stressors that they are willing to deal with and those they won't. I came from a corporate environment where I worked for fifteen years in a day to day job that was predictable (more or less), and I did what I was told. I dealt with corporate executives who made millions of dollars treating people in an abusive fashion. For me, working for myself and having the flexibility of being able to say when I work where, how many hours I work at each job, is less stressful than the corporate nine to five with all the financial perks. Also, I prefer the unpredictability of my job. I have clients with a myriad of disorders, and am exposed to things that private practice would never expose me to.

Does that mean that I love my work every day? No. Some days I'm frustrated by the limitations. Some days I'm frustrated by the system. Some days I'm tired of the abuse the kids hurl at me. But for the most part, the good days out weigh the bad, and I usually look forward to Monday and finding out how my kids' weekend went. Every fall when the new school year starts I look forward to seeing how my kids—especially the boys—have grown. I get to watch boys who start off shorter than me literally grow into young men. I watch as they go through adolescence and the awkward stages associated with puberty.

And when a case comes along like Andy, and I get to see someone come into their own and leave my school for a better future, I know that all the frustrations, limitations, and abuse has been worthwhile.

2

A Typical Client

So what does a typical "hard-to-reach" client look like? Quite honestly, it will depend on the setting. When I worked at a Continuation High School, the students there were higher functioning. They were academically capable and did not have learning disabilities—they were simply behind due to truancy issues or having missed school due to pregnancy or time in Juvenile Hall or Camp (and I don't mean Camp Hiawatha with songs and marshmallows). These were kids who all lived with their family of origin, and for the most part, they suffered from depressive or anxiety disorders. Some had anger issues—but I do not recall any diagnoses, such as ADHD, that would have required psychiatric and medication intervention. These were kids who simply needed someone to talk to.

The client I remember most from this school was a young girl, Valerie, who suffered from Turner's Syndrome. This is a genetic disorder that causes a girl to miss one of the two X chromosomes. As a result, the girl's ovaries do not develop normally during puberty. Therefore, Valerie did not menstruate or develop breasts. There are also several other physical characteristics that come with the disorder, such as short stature, as well as cognitive ones—and people with Turner's Syndrome can vary in the degree to which they have these characteristics. Visually, you could not tell anything was wrong with Valerie, other than she was flat-chested. However, in terms of her self-esteem, the older she got, listening to other girls talk about their period, the more depressed she became. No one had bothered to educate Valerie on her condition.

When I met Valerie, I had never heard of Turner's Syndrome, so I researched it, and brought in the information from the internet. When we started working together, she was seventeen. She confided in me that the other girls thought she was weird, because they would ask her when her period was, and she would make up a date. Knowing nothing about menstrual cycles, her stories obviously didn't add up. As I worked with Valerie, she became more confident in herself, and took the material home asking her mother to take her to the doctor for the necessary hormone injections, which they resumed. While medically it was too late to repair the physical damage, psychologically Valerie made progress. Her grades went from near failing to Cs and even a few As. The administrative staff told me that Valerie was like a different person on campus—she smiled and laughed, and had started making friends instead of isolating herself.

When I worked at the school in Watts, the students there were predominantly black, Hispanic, or mixed. Interestingly, the ones of mixed ethnicity proclaimed they were black, because it was "safer." The population at this school was primarily learning disabled, ADHD, oppositional defiant disorder, and there were very few severe psychiatric disorders. The few kids with outright behavioral problems—anger management, conduct disorder, or physical aggression—usually ended up being sent to the Halls after less than twelve weeks, if even that. There were also a lot of gay students at this school, as well as gender identity disorders. However, due to the raging homophobia that existed in this particular school, none of the kids ever admitted to their sexual orientation. I think this had nothing to due with a lack of trust in me, but on the part of the staff. The staff at the school was almost all black (except the counselors who were all white), and made it clear how they felt about homosexuality.

There was one classroom at the school in Watts devoted to Autism (taught by the only white teacher), and I did work with a lot of the autistic kids because I had expressed an interest in learning about the disorder. There were also a lot of kids who had mental retardation. While there did not appear to be many kids with drug problems—at least none of my students were ever caught loaded at school—there were a lot of problems with students who had been born with fetal-alcohol syndrome or some similar at-birth cognitive impairment due to the mother's drug usage.

I worked at Watts when I was an intern, and I worked full time. At the time, I thought the work was difficult. In retrospect, it was easy because most of the kids in the school had only one disorder—there were very few dual-diagnoses. I doubt if there were any multiple diagnoses.

The group home where I worked for two years had a mix of ethnic groups—black, Hispanic, and white. The major problems there were ADHD, anger management, drugs, drugs, and more drugs. These were the type of kids who would drink cough syrup to get loaded. I had one client who would repeatedly fall asleep in session he was so stoned. The staff would make repeated sweeps of the kids' rooms, but when kids are able to buy their supply at school, what can you do? They eventually started stripping the worst offenders when they walked in the door every day from school.

Almost all of the kids in the group home suffered from oppositional defiant disorder. I think it was the standard diagnosis given when a new resident walked in the door. Think of having a perpetual two year-old in the house (or six toddlers, as was the case). You tell the boys it is time to do their home work, they say "no." It's time to shower before bed—"no." It's time to go to bed—"no." These kids were programmed to say "no" to any adult request. When working in any school or facility, kids go through a honeymoon period where they behave well until they assimilate. Once they feel comfortable in their surroundings, their "true self" is revealed. This is true in a group home and a school. Whenever a new kid arrived at a group home, a kid would help out the staff, do what he was told, while the other kids would call him an "ass-kisser," and other similar names. But for the most part, the most common heard word at the group home was "no." Of course, it went both ways. When the kid would ask the staff if they could have a home pass, the staff would look in the record book at how many marks they had received during the week, and respond, "no."

When I started working at my school in the San Fernando Valley, I began working with disorders from the DSM-IV (Diagnostic and Statistical Manual of Mental Disorders) that I had never encountered before. Most of the kids on my caseload have multiple diagnoses. A simple dual-diagnosis would be a welcome relief. I have predominantly blacks and Hispanics, but also a few Asians, whites, and sometimes I'll get a different ethnic group that does not fit the standard norm. I have kids who range from low-functioning—meaning they will probably grow up to live in assisted living housing or stay with their family of origin for life, to high-functioning kids who pass the high-school exit exam, work jobs part-time after school and during the summer, and go on to college.

In terms of disorders (and this is why I realized my internship at Watts was a walk in the park in retrospect), I have worked with Autism, Asperger's, ADHD, bi-polar disorder, conduct disorder, depressive disorders, encopresis, mental retardation, schizophrenia, social phobia, and the run of the mill oppositional defiant disorder and learning disorders. Usually students have a learning disorder, plus an emotional disorder—or two.

An example of one of my lower-functioning clients was a young boy named Billy. Billy was approximately eight when he was added to my caseload, although it was hard to tell because he was approximately fifty pounds overweight. Billy was the youngest male in his family, and had been spoiled. He apparently got whatever he wanted at home, and while he was academically capable of doing his work, he was unwilling to do anything asked of him because he had learned to tantrum to get his way. He suffered from anxiety, possibly due to his fear of coming to school and being teased. I never could really tell where his anxiety stemmed from because Billy lied about everything.

Billy would bring his homework in every day done perfectly— but it had clearly been done by his older brothers. When you asked Billy to repeat a math problem, he would shut down and become defiant. When the class had lunch, he would eat his dessert and refuse to eat his lunch, saying that chocolate cake was healthy, but vegetables were bad for him.

Billy suffered from trichotillomania (hair pulling), and started to get a bald spot on his head where he constantly pulled his hair when anxious. He would also pick at his arms and develop scabs. While it was easy to feel sorry for Billy, due to his constant state of anxiety, he would never let any staff in emotionally to discover where his anxiety came from. When you confronted him on his weight issues, he gave such convoluted explanations—such as how chocolate was good for him—that was almost insane (it should be noted that both of Billy's parents were overweight).

Billy literally turned into a lump in his class. When it was time to do class work he would sit at his desk, arms dangling to his sides, refusing to comply. He would sit and stare into space, or repeatedly kick the desk in front of him. However, when it was time to interact with other children, Billy would suddenly become a passive-aggressive bully.

While sitting in circle-time, he would quickly kick anyone who was near him, without any provocation. At physical education, he would hit the other students with a ball on purpose. The older, and larger, Billy got, the more aggressive his behavior became.

A little over two years after being on my caseload, without any signs of success, Billy had developed truancy issues. He rarely came to school, and when he did, he spent his entire time at school screaming at the top of his lungs "I want to go home," over and over, until he would have to be removed from the class. He would kick the walls, overturn desks, and physically attack anyone who came near him. His parents were continuing to give him whatever he wanted at home, and he thought the same tactics would work at school if he threw a tantrum long enough.

One day, it was an hour after school had started and I was crossing the grounds. I saw a car parked in the driveway which was rocking back and forth. I could see that in the car was a child out of control. I witnessed Billy, assaulting his mother, punching her in the face, while the father sat in the front of the car and did nothing. Through the closed doors of the car I could hear him screaming, "You fucking bitch! You can't make me! I hate you, you fucking bitch! I want to go home!" I stood there and stared, waiting for the father to do something to stop his child from assaulting his wife. Eventually, two aides from the school removed Billy from the car, but the entire time the father sat passively in the car. Unfortunately, when parents grow up with this behavior, sometimes they simply give up. No doubt what looks like a horrific case of abuse to an outsider, is just another normal day for the family.

In sessions with me, most of Billy's play had been about death and violence. Not about suicidality, but about killing family members, especially his younger baby sibling. Yet nothing in Billy's profile fit that he had been abused in any way either sexually or emotionally. Certainly not physically. My only concern for physical abuse was for his family.

To make a long story short, it turns out that Billy was supposed to be on medication all along, but the parents did not want Billy on the medication, because it caused him to gain weight (apparently medication to control physical aggression is bad, but chocolate cake and an endless supply of Doritos is not). I filed a child abuse report, and it was eventually recommended that Billy, who was a danger to his family, be removed from the home to an environment where his medication would be properly regulated and monitored.

One of my higher-functioning clients was a boy named Jared. He came to our school from a public school, and this was his first non-public school placement. He had been expelled from his home school for physical aggression and peer-relationship problems. Jared had anger management issues, but otherwise seemed developmentally fine. He was academically on track, and had no ADHD or other disorders. He was just angry. He came from a broken home, and seemed to have issues with his stepfather, or rather, issues resulting from the lack of contact with his biological father.

In therapy, Jared was polite, but closed off. In his first session with me, he was polite and cooperative, and answered all of my questions. Then on the way back to class he boasted, "At my last school I killed a teacher." I just said, "Wow, it sounds like you want me to believe you're a tough little banger." He stuffed his hands in his pants and said, "You got that right, miss." Jared was eleven, and through out all of our time together he would refer to me by "miss."

I taught Jared how to play Texas Hold 'Em—not something I would normally teach someone so young, but Jared was very smart. He loved it, and started always placing the same bet—forty-one chips. That became his shout out when he saw me on campus—"forty-one!"

Jared was good at sports, and joined the school's various teams. While he still continued to be somewhat emotionally guarded in session, he came every week and participated for his full time. When he saw me on campus he always smiled. At his IEP his parents complained that he didn't do his chores or respect their rules. We talked about this in session, and he listened, understanding what I said, but never commenting. It was like Jared wasn't going to give me the satisfaction of letting me see him get emotional or having an actual dialogue.

However, when I spoke to Jared's teachers and aides, the reports were always positive. Jared was helpful in the class, and tended to help the kids who were picked on—those who were developmentally disabled or disadvantaged. He never instigated the autistic or mentally retarded children as so many of his peers did. When playing games, Jared would deliberately let them win.

Jared was eventually returned to public school after eighteen months in a special education setting. We spent the last month together talking about how things had changed for him at home, and why he had been so angry. We discussed what he would do if he got angry at his new middle school. I teased Jared that one day I would show up at his home school to shout "forty-one" behind his back when he was walking down the halls and he least expected it. He gave me a cocky grin, and said, "Go ahead, you do that miss. See what happens."

In many ways, Jared had not changed. He was still full of bravado. I never knew what he got out of counseling, but that's the way it is with many of these kids. The changes are subtle, and when you see the clients five days a week on campus, you can't pinpoint when the change occurred or what exactly the change was. In the end, it does not matter. You do your job and move on. And hopefully, so do your clients.

3

Bangers Have Feelings Too

Unless you work with the Amish or Mormons, chances are if you work with adolescents, you will have some gang-affiliated clients. I have worked with gang members, or children who live in a gang territory, since I started working in the school system in 2001. It's easy to spot the bangers—the kids who boast out loud in class that they are gang-affiliated are not; the ones who keep it quiet (on the down-low) are. Gang members are expected to keep a low profile in the schools due to the very fact that schools, theoretically, are not supposed to have gang-members enrolled (there are locked-down schools exclusively for bangers).

Almost all of my kids at my various schools who have had gang affiliations let me know about their membership as soon as trust was established. They usually tell me with eyes downcast and with shame. Banging is a way of life for these kids, and it is not something they are proud of. Most gang members come from a one-parent family, have one parent in jail or deceased, and have multiple (male) family members who were in gangs and/or were killed through gang-related activities. Jumping in, as initiating is called, is almost a rite of passage.

Avoiding the life of a gang member can be almost impossible for these kids. I remember at the school I worked at in Watts, when a young black man said to me in a dejected tone, "I hate my neighborhood. I can't sit out after 6:00PM. I can't walk around without the cops pulling me over. I can't walk down the street without worrying about getting jumped. My mom's working two jobs so we can move to a better neighborhood, but even then I'm scared they're (the gang) still going to find me." At barely sixteen, this young man felt his fate was sealed because of the area he was born into.

When working with gang bangers, it is important to have a non-judgmental mentality. These kids have no more control over their destiny than children have over the religion their parents pick for them. While I am by no way implying that gang-banging is similar to religion, what I am implying is that it is a lifestyle that children are born into and they have no control over. It is doubtful that a therapist would judge a child for being Jewish, Catholic, or Baptist, simply because their parents are of that denomination.

When a child is old enough to go to a religious service for the first time, parents do not sit down and say, "Well, son, your mother and I are both Roman Catholics, but if you want to be a different faith, that's all right with us. We'll take you to a different church every week so that you can experience different religions, and when you make up your mind, you pick what you want to be." It is assumed that children will be the same faith that their parents are. Even then, when a child grows up and leaves a church that the family attends, or marries outside that faith, there can be family discord and upset. This is how it is for families who have been in a gang for generations. It is assumed that the children will be in the same gang.

I was assigned a student who informed me he missed the second and fourth grades, which was why he was behind a grade. I asked him what he was doing when he was truant at age seven. "Banging," he replied. When I expressed surprise that he had jumped in so early, he shook his head and corrected me. He had joined in at age five. He shrugged, and told me that any kid of his would be born into the same gang as soon as it came out of its mother's belly.

I worked with a young female, Maria, for many years. She had belonged to a gang, but denied being currently active—a difficult thing to do, since once you joined it was not technically possible to leave. She had tried many times to eradicate her tattoo that showed her gang-affiliation to no success. However, Maria was a model student, came to school fairly regularly, and tried her best. In counseling, she was open and honest about her insecurities with her academics and talked about her problems at home—of which there were several.

Maria was a sensitive, deep soul. Her mother was an alcoholic, who never complied with any treatment recommendations. Maria had several older brothers who were in jail. She had a younger sister who rarely came to school. Eventually, Maria's sister would end up in Juvenile Hall, and later return to school sporting an ankle bracelet, only to disappear completely. However, Maria took responsibility for her sister's failures, and much of my work with Maria was to try to get her to focus on healing herself, and not to worry about fixing everyone around her.

Every year around the holidays Maria would come to school in tears, requesting to see me as yet another friend or family member she knew would have died in a gang-related shooting. A phenomenal artist, Maria would use her sessions to draw pictures of her dead acquaintances.

Grief and loss work is a common theme with my gang members. I had one fifteen year-old male who would discuss a friend or friend of a friend who had been killed every three or four months. Brian had an "I don't care" attitude about everything, and denied being worried about the consequences of his gang activity. He had major truancy problems, and one day when he had not been at school, I actually saw him in a truck with his gang buddies in another part of Los Angeles in the evening when I was attending a Continuing Education Course. The following week, I asked him, "Hey, last Thursday, were you in Whittier with your crew?" He looked at me, scared. "I knew it! I told my homies some white chick was staring at me and they said I was paranoid."

I often joke with my students that no matter where they are and what they are doing, I'll find out about it. I reminded him of this, and then asked him, "Did your mom know where you were? Cuz you weren't in school that day." Brian looked sheepish, and begged me not to call his mother. It is easy to forget when dealing with teens who talk about tagging, cruising, and stealing, that they are ultimately *children*.

One of the most frustrating cases I ever worked with was a young man named Jesus, who came onto my caseload at age fourteen. Jesus was not in a gang when I started working with him, but all the males in his family had gang affiliations, and many were dead or in jail. Jesus lived with extended family members, since both his parents were dead. Like Maria, Jesus was not in the system. He had no outside social worker or therapist to talk to, and I was the only person he had who would sit and listen to him each week, giving him my total and undivided attention. He used his session time wisely, talking about his feelings about his family, his problems in school, and his fears about his future.

Jesus lived with his two aunts, and therefore had no male family members to bond with. He seemed to have trouble opening up to the teachers at the school, perhaps because of the racial differences, but probably because of the fact that his teachers changed every year. I watched as Jesus became more and more depressed, and his academics suffered as a result. Cognitively, Jesus was bright enough to earn high marks in school—however, he lacked the motivation. He preferred to play the class clown and get attention by being the "smart-ass." Every year in his IEP I told his family that if Jesus worked as hard in class as he did in counseling, he would be earning B's.

Jesus starting talking to me about suicide, but denied that he would ever harm himself. He was religious, and felt to take his life would be a sin. Shortly after the topic of death came up, Jesus confided in me that he was being pressured to jump in to the same gang his family had belonged to.

"Have you already joined?" I asked with raised eyebrows.

Looking down at the floor, Jesus nodded.

Jesus and I spent the rest of the hour talking about how just maybe his joining a gang was his way of killing himself without having to pull the trigger himself. I told him, "If you join a gang, it's the same as if you put yourself in a casket, cuz sooner or later you're gonna get shot. I don't think this is about you wanting to bond with your family. I think this is about you being so depressed you don't care if you live or die anymore." As usual, this conversation was right before the Christmas break, when most "emotional" conversations take place for the students. I talked with Jesus about where he could go over the winter break to get free counseling while not in school, and we talked about ways to stay safe.

During our three years together, Jesus would disappear for months at a time, during which he would be living on the streets, only to reappear again. He confessed to committing a myriad of gang-related crimes while living on the streets, and doing assorted drugs ranging from marijuana to crystal meth. He also confided that he was tired of living in continual fear that he would look over his shoulder and find a patrol car behind him.

Sadly enough, Jesus, like many other bangers, would ultimately abandon his education for life on the streets. As he said to me one day, "I think things would be easier if I just got picked up and spent the rest of my life in jail." For a lot of children with low self-esteem issues, depression, and feelings of alienation—living the regimented and structured life of a prison sentence is far easier than living in the real world where the fear of failure lurks around every corner. Jesus, at age eighteen, quickly disappeared from school, and I never learned what his fate was.

One of the most at-risk targets for gang-related criminal activity are teens with cognitive disabilities. These adolescents, with borderline intelligence, are so desperate to fit in with their peers, they are often targeted by gangs to perform criminal acts—usually theft or murder. One of the students I worked with for two years, Enrique, was such a student. Enrique was basically a good kid. While biologically seventeen, he was cognitively around six. He was unable to perform math or English beyond the first grade level, was unable to tell time, or read a calendar. However, Enrique understood right from wrong, and was willing to stand up for himself.

While the school was encouraging Enrique's mother to seek legal conservatorship before he turned eighteen, Enrique's friends had other plans for him. Repeatedly he would come to school with his face bruised or bloodied, and when I questioned him, he would tell me a story similar to the following. His neighborhood friends would try to give him a gun, and tell him to shoot someone, and he would refuse, and then there would be a fight. At one point, Enrique disappeared from school for two months. He had been caught with his friends in a stolen car. Luckily, the cop who caught my client was able to ascertain that it was impossible for Enrique to have stolen the vehicle, and did not press charges against him. When I asked Enrique why he was friends with people who continually tried to get him into trouble, he stated, "They buy me things."

While Enrique understood that the consequences of shooting and stealing could get him in jail, and he denied being involved personally in these activities, he was unable to understand how associating himself with those who regularly committed such crimes was endangering his own future. Enrique just desperately wanted to have friends, and in his limited cognitive capacity, corrupt friends were better than no friends.

Not all gang-affiliated clients have such negative outcomes. I worked with a young man, who at age sixteen, came to me while still residing in a group home. He was on probation for a minor charge. As was often the case, he had been in the wrong place with the wrong crowd. Quentin had a slight cognitive impairment. While he could academically grasp his school material, he needed significant help, and more basic life skills, such as the months of the year and telling time eluded him. He once missed his probation hearing because of being unable to tell what day of the week it was or what month it was (in psychiatric terms the basic oriented times three).

After working with Quentin for several months, he let me know he used to be in a gang. He started tagging on the white-erase board in my office, and showed me the various crews he used to hang out with. At one point he wrote out a phrase with the word "CRIP," which I took and changed to emphasize the "RIP."

C R I P   ⟮C R. I. P.⟯

"Funny thing about gangs," I said, "Is that if you belong to one, you better be prepared to rest in peace."

Quentin burst out laughing, thinking it was a joke, but then he studied the board a little longer, and cut his laughter off. "Damn," he said, "I never thought about it like that. R.I.P. You got that straight."

Quentin eventually returned to his family of origin, but remained on probation. While at school, he joined the choir, the basketball and football teams, and worked in the school store. In choir he excelled— he was naturally outgoing and took direction well. The staff could count on him to help out, and he was well liked by his peers who saw him as non-threatening. Quentin treated all the kids at the school with respect, even the autistic and mentally challenged children—often high-fiving them, or helping them when he passed them on the grounds. He never discriminated based on race or disability.

While Quentin struggled academically, especially with his reading which we often worked on in session, Quentin passed the math part of his high school exit exam. Quentin wasn't perfect—he changed his career goals on a monthly basis, and even though he was going to be eighteen soon, he still wanted others to buy things for him rather than supporting himself. He spent his free time out on the streets partying rather than looking for a job. But he tried hard to improve himself, and to leave his gang history behind. Quentin was chosen to give the culmination speech at his graduation, which he wrote in a session in my office. He was proof that you didn't have to let your past dictate your future.

4

Just the Facts Ma'am

So where exactly do these children come from? As I mentioned in the previous chapter, it depends on the setting. Different schools get referrals from different resources. At the school in Watts, despite being in a lower socio-economic neighborhood, literally in the middle of two rival gang factions, the kids were actually higher-functioning academically. I only had one bi-polar child on my caseload the entire time, and she was actually a very intelligent child (she was in the music business). At my school in the San Fernando Valley, where gang activity is pretty non-existent, or at least, if gangs are present, they are keeping themselves hidden, the kids at the school get asked on almost a daily basis, "Did you take your meds today?" Over ten percent of my caseload is psychotic alone.

So again, where do these kids come from, and what does their future look like in terms of statistics? As I said, many of the kids at both the SED schools I have worked at are in foster care. But what does this mean? Foster care can mean a child lives in a family home (either relative or non-relative), a group home, or some other sort of placement. Many of my students live with relatives who are being paid by the state to serve as foster families. This is usually because one parent is dead, and the other parent is in jail or deemed unfit. In some cases, due to abuse issues, the legal parents have lost parental rights to their children. These children are usually black or Hispanic. But again, this is based on my current school in one location in California. Foster care composition by state varies greatly. According to the Adoption and Foster Care Analysis and Reporting System, there were an estimated 510,000 children in foster care in 2006. The average age of these children was 9.8 years.

Placement Settings of Children In Foster Care
September 2006

Placement	Percent	Number
Pre-Adoptive Home	3%	17,351
Foster Family (Relative)	24%	124,571
Foster Family (Non-Relative)	46%	236,911
Group Home	7%	33,433
Institution	10%	53,042
Other	10%	44,691

Source: The Adoption and Foster Care Analysis and Reporting System, 10/1/05 through 9/30/06

Nationally, boys (52%) constitute slightly more than girls (48%) in foster care. Whites (40%), blacks (32%), and Hispanics (19%) comprise the majority of the ethnic groups in foster care. Approximately 53% of the cases exiting foster care in 2006 (or 30% of all foster care cases in total) were reunified with their parents or primary caretaker. However, a shocking 509 cases resulted in death that was attributable to medical conditions, accidents or homicide. The average length of stay in foster care is slightly over 2 years (28.3 months).

Average Length of Stay In Foster Care
September 2006

Less than One Year	42%
12 to 23 Months	22%
2 to 4 Years	23%
5 Years or More	13%

Source: The Adoption and Foster Care Analysis and Reporting System, 10/1/05 through 9/30/06

California (14%), Texas (6%), Pennsylvania (5%), New York (4%) and Ohio (4%) had the highest level of children entering foster care in 2006, and these levels have remained fairly consistent since 2002. The one exception was Texas, which in 2002 ranked sixth, but due to a steady increase in the number of children in foster care over the years now ranks second. The states with the lowest level of foster care children—their levels are so low they do not even compute to a percentage—are New Hampshire, the District of Columbia, Vermont, and Maine.

In terms of the national population of students (5,867,234) being served through the Implementation of the Individuals with Disabilities Education Act (IDEA), the majority of these students have a specific learning disability (49%), followed by a speech or language impairment (18.6%), or mental retardation (10.3%). Blacks are more likely to have mental retardation (17.4%) or an emotional disturbance (11.3%) than the national average, while Hispanics are more likely to have a learning disability (58.9%). Autism appears relatively the same across all the races sampled according to the U.S. Department of Education, with the exception of a slight spike among Asians (however, sample sizes were not available to know if this represents a statistical significant difference).

Disability Distribution of Students ages 6-21
Served Under IDEA (2001)

Disability	Asian	Black	Hispanic	White	All Students
Learning disability	42.1%	45.4%	58.9%	48.1%	49.2%
Speech or language impairment	25.1	14.6	17.7	20.0	18.6
Mental retardation	9.4	17.4	8.1	8.6	10.3
Emotional disturbance	5.0	11.3	5.0	8.0	8.1
Multiple disabilities	2.7	2.1	2.0	2.2	2.2
Other health impairments	4.4	4.3	3.2	7.0	5.8
Autism	4.1	1.4	1.1	1.8	1.7

Source: 25th Annual Report to Congress on the Implementation of the Individuals with Disabilities Education Act, Volume I, 2003.

One thing I have learned about student disability classifications after so many years of participating in IEPs—they are often misrepresented. So the data in the above chart, I personally take with a grain of salt. I know for a fact that students who do not have learning disabilities, or emotional disturbances, will keep these classifications in their IEPs in order to get specific services when they continue in public school or college. I have had numerous children come in with parents demanding speech services, when their child speaks better English than many of the staff at my school. I often get an IEP for a student with a classification of learning disability, but no classification of emotional disturbance, or vice versa, when the child clearly fits the criteria of both. So I suspect the number of students with multiple disabilities is higher. On the other hand, it could just be a reflection of the schools I worked at and the fact that I'm in California where the system tends to be overloaded. Maybe if I were in Vermont...

There has been a lot of coverage about Autism in the news in the past few years. Parents of autistic children are on talk shows. Specials air on the news channels and even HBO. There are controversies about the causes and the cures. According to the U.S. Department of Education, the number of children ages 6–17 with Autism served under IDEA has jumped from approximately 15,000 to 90,000 from 1992 to 2001. There can be several reasons for this increase in numbers. Depending on how young children are diagnosed, other pervasive developmental disorders or speech impairments can mimic Autism. As I mentioned earlier, once a disability is put into the computer in the student's IEP, it is often left unchanged. Many parents can't be bothered to show up for their child's IEP, either from lack of interest, or they are unwilling to take off from work. In the case of many foster parents—they simply do not care. They aren't *paid* to attend school meetings. Disability categories cannot be changed without the consent of the parent or legal guardian.

In addition, the field of psychiatry is relatively young. Over one hundred years ago, a schizophrenic would have been labeled as "insane" and locked away in an asylum. Did that mean schizophrenia did not exist? Of course not. The term schizophrenia did not even come into vernacular until the early 1900s and its current definition bears no resemblance to its earlier version. We now have disorders such as schizophrenia, schizoaffective disorder, brief psychotic episode, manic episodes, and several other diagnoses that stem from the symptoms associated with the original term "schizophrenia" (Campbell, 1981). The point is, in the early

1900s, if a person was seen wandering around the streets talking to themselves, they would be deemed "demented" or "possessed."

Autism has undergone similar fine-tuning in terms of understanding and diagnosis. A child in the 1950s who had Autism would no doubt have been diagnosed as mentally retarded, or having seizures or "fits." As the population at large becomes more aware of a diagnosis, parents who have a child who fits the criteria is more likely to seek professional help, which drives the numbers higher. Fifty years ago, bizarre mannerisms such as a child who was mute and flapped their hands would have been kept locked at home—away from prying outside eyes. Today, thanks to media awareness, parents do not feel shame and seek immediate treatment. Does this mean Autism did not exist in the 1950s? Not necessarily. Unfortunately, it is a question that cannot ever be answered.

However, it has been my experience that many students labeled with Autism do not in fact have the disorder. I had a student who graduated from school—he had passed his high school exit exam and was going on to college. While at our school he had played sports on the various teams, worked in the various work settings, and was trusted to help out the staff. In his exit IEP he asked about the eligibility category in his IEP, which was Autism. He wanted to know why it was there and wanted it changed. He was concerned about this label following him to college. Clearly, he was not autistic. When asked, he said he thought he had been diagnosed around the age of three—he was now seventeen. He said at that time he was very shy and rarely spoke—plus he mumbled when he spoke so people didn't understand him. No one had bothered to correct the mistake along the years. Unfortunately, his parents did not speak English, and would probably never take the time to change this paperwork error.

But probably the most compelling story, and most common, was with a young boy named Manuel. Manuel was in adoptive placement, as was his little brother who also attended our school. Both were diagnosed autistic. Manuel's strongest asset was his social skills, which is contrary to the diagnostic code for an autistic disorder (which I will address more in depth in the chapter on working with developmental disorders).

Normally, when a parent is informed that their child has Autism, the reaction is one of shock, disbelief, and dismay.

They may demand a second opinion. In Manuel's IEP his adoptive mother kept talking about how she and her husband had structured their home to correct Manuel's autistic behaviors. I stopped her and said, "You keep referring to Manuel as autistic. Manuel is *not* autistic."

Instead of looking relieved, she looked angry, and said sharply, "Why is it that all the therapists and teachers at all the schools keep telling me my son isn't autistic!"

I gave her a questioning look. "You don't seem relieved. I'm surprised."

"I took him to a very expensive psychiatrist who gave me a diagnosis of Autism!" She then went on to demand to know what my qualifications were. Of course, the key phrase was, she went to a doctor who "gave her" what she wanted. It was at that moment I remembered that this was an adoptive—or foster—parent. Foster parents get paid more money from the government if the children they take care of fall into the autistic category. Therefore, by fostering two autistic children, these people were getting even more money.

One parent of a child with multiple disabilities flat out told me she lied about her son's Autism to get additional services for her child; however, her son was actually seriously impaired and was a danger to himself and others. This was the only way she was able to get the services she needed in the home, such as respite care. This suggests that there is a problem with the system. One thing is clear—the numbers alone can't be trusted.

Another number that is probably also over-reported, is the percentage of kids who are on psychotropic medication. This information is from a parent survey, and it is my experience that most kids cheek their meds. When I worked in the group home, medications would be found under mattresses, or in pillow cases. Kids would go into the bathroom and make themselves vomit after getting their medication. So this information should be considered to reflect the number of children *prescribed* medication; not necessarily those *taking* medication.

Use of Psychotropic Medication Among Children With Disabilities
2000–2001

	6–9 Years	10–12 Years	13–14 Years	15–17 Years
Any Psych Medication	17%	22%	21%	18%
Stimulants	14%	18%	18%	11%
Anti-Depressants/Anxiety	5%	6%	8%	9%
Other Psychotropics	4%	3%	10%	11%

Source: 25th Annual Report to Congress on the Implementation of the Individuals with Disabilities Education Act, Volume I, 2003.

The majority of the children who are taking anti-depressants or anti-anxiety medications tend to be children who are classified as having emotional disturbances between the ages of 6–12 (40%) and 13–17 (29%) or other health impairments (47% and 38% for 6–12 and 13–17, respectively). Emotional disturbances comprise the highest classification of children receiving anti-psychotic medication with almost an equal proportion among both 6–12 (24%) and 13–17 (29%) year olds. The overall disability group which was least likely to be on psychotropic medication was speech/language impairment (approximately 9% ages 6–17). The one classification with the largest discrepancy from 6–12 year olds (3%) and 13–17 year olds (43%) in terms of use of psychotropic medication was Autism. There is one possible explanation for this—the triggering of hormones during puberty around age eleven causes an increase in many autistic behaviors, including self-injurious ones—and no doubt many parents who are reluctant to put their children on medication when they are little do so once their child's behavior becomes more out of control and self-harmful.

In terms of behavioral problems among students with disabilities, 27% of kids 10–12 year of age reported having been physically attacked or involved in a fight in 2000–2001. This level was fairly constant among the 13–14 year olds (25%), 15 year-olds (23%) and 17 year-olds (22%). It should be noted that this information is again based on a parent survey, and the actual percentages are probably higher. My experience is many parents are clueless as to what their children are doing in school. When a child gets into a physical altercation at our school, an Incident Report has to be sent home. But often, the staff does not want to bother with the paperwork, and if no one was injured they do not fill out the report. Also, as I said, foster parents do not often care, and do not read the pieces of paper that come home from school.

One of my clients, along with a group of other students, was suspended for assaulting a new staff on the bus who as a result of the assault ended up hospitalized. None of these students could return to school until their parent or guardian came to the school to talk with the principal. The principal of the school asked me to join this particular student, Allan, when we met with his mother. Allan's behavior over the past two years had escalated out of control. The school had held several FAA/BIP (Functional Analysis/Behavior Intervention Plan) IEPs for him when he had assaulted staff or destroyed property. His mother had never attended even one of these IEPs. For Allan to return to school, his mother now *had* to come to the school.

Allan's mother was indignant that the school had not notified her that her son's behavior was so out of control. I spoke up, "Mrs. Smith, we have had over three IEPs for Allan in the last year due to his behavior problems, and you have not attended one of those meetings. You signed the releases, allowing us to conduct the IEPs in your absence. The reports were sent home to you. If you had read the reports, you would know that your son's behavior had deteriorated in the past two years. I hope, given this suspension, you'll be able to find the time to attend his next IEP." Later in session, Allan said to me, "She don't read nothing you guys send home. She tosses them in the trash!" Allan did not say this in a defiant way, as if to say—ha, ha, I can get away with anything. Allan was angry, because his mother cared so little about what he did. His behavior was clearly an attempt to get her attention.

Nearly one-third of secondary school-age children with disabilities were reported by their parents to have been suspended or expelled from school. Among elementary and middle school children ages 6–12, blacks (28%) are more likely to get suspended or expelled than either Hispanic (13%) or white (10%) students.

Suspension or Expulsion Of Students with Disabilities in 2001

Suspended or Expelled	Age 13–14	Age 15	Age 16	Age 17	Total
Percentage	27.3%	34.1%	35.5%	35.7%	32.7%
Number in Sample	3,021	2,194	2,215	1,410	8,840

Source: 25th Annual Report to Congress on the Implementation of the Individuals with Disabilities Education Act, Volume I, 2003.

In terms of criminal activity handled by the juvenile court, 556,500 juveniles were sentenced to probation in 2005 according to the Juvenile Court Statistics Handbook. Probation is defined as the decision of the court to have the youth placed on either informal or formal, or voluntary or court-ordered, supervision. This represents 33% of all the cases handled by the juvenile court system. Of the cases sentenced to probation, 74% were male compared to 26% female. One-half (50%) were between the ages of 13–15, and another one-quarter (24%) were 16 years of age.

An additional 140,100 youths were placed into a residential facility for delinquents or otherwise removed from their home of residence. As with probation, males (83%) outnumbered females (17%) in terms of the percentage of youths put into placement. Teens 13–15 (47%) were followed by 16 year-olds (26%), and 17 and over (23%).

So what are the crimes these adolescents are committing to get themselves in trouble? In terms of total arrests, including adults, juveniles were responsible for 49% of arson, 39% of vandalism, 28% of robbery, 28% of burglary, 25% of motor vehicle theft, 24% of weapons, 14% of aggravated assault, 10% of drug abuse violations, and 10% of murder (Office of Juvenile Justice and Delinquency Prevention). Blacks represent a significantly higher proportion of the juvenile arrests for robbery (67%), murder (59%), motor vehicle theft (43%) and aggravated assault (42%). Unfortunately, Hispanics are considered an ethnic group and included with whites, so there is no break out for this minority group for comparison purposes.

Hopefully, in the future the government will track juvenile delinquent behavior based on information beyond age and race. Given the proportion of the students I have worked with who have learning disabilities and/or emotional disturbances who are also on probation, perhaps the government would be more motivated to spend money on early intervention and education if there was a clear correlation between low self-esteem due to problems in school and in the home, and gang-related and/or criminal activities.

And what about the gangs? Statistics for them exist to some degree. But most of the statistics I found were only available on a city-by-city basis—and all claim to be under-represented. Because I could not find any reliable source of information, I decided to forgo reporting gang numbers.

The point I'm trying to make with all the numbers included in this chapter is this—there is a serious demand for people who are willing to work with this population, and this is just a snapshot of what these kids' lives look like. In some way, their lives are like these numbers. When they get turned over to a caseworker, they get assigned a number, and they literally become a statistic in the system.

5

First Contact

When assigned a child in a placement setting, they come with a file of one type or another. These files can be up to several inches thick. I tend not to read it until after I've met the child. I'll skim the file looking for red-flag words such as "suicidal," "psychotic," or "sexual abuse," but I prefer to let the client tell me their story in their own words, so that I don't prejudge them. I learned this when I was an intern. While working in Watts, I was assigned an autistic child named Jose who had come from another non-public school. He was approximately ten or eleven. His IEP read like a nightmare. One thing was clear—his previous therapist did *not* like Jose.

The therapist stated in Jose's IEP that Jose could never benefit from counseling, was not appropriate for an open-school setting, and he went on to basically compare Jose to a wild animal that was untreatable. Jose apparently had thrown fecal matter at his therapist. Needless to say, I was a little hesitant to get Jose for his first session. I asked his one-on-one to come into my office just in case. She sat nearby in a chair while Jose sat on the floor. I had an old Playschool castle which had survived my childhood that Jose immediately gravitated toward. It turned out that Jose was mid-level functioning, and was easily engaged on the social level. The first session went well, and that was the only session that I had his aide attend.

While Jose was playing with the castle, a character fell into the moat, and I said, "Uh-oh, be careful, or the alligator's gonna bite him." That phrase, "The alligator's gonna bite," would be the line Jose would use to connect with me frequently, and I would catch him deliberately drop the Playschool characters into the moat to illicit this response. As our relationship grew, we would play peek-a-boo through the castle windows, and later he would play peek-a-boo behind his shirt or jacket when he saw me on the grounds. Jose had huge brown eyes, and I nicknamed him "brown-eyes," which would make him smile and light up the room. Jose was overweight, as are many autistic kids. Jose weighed at least fifty pounds, if not more, than I did, and would walk down the long school hallway to my office with his arm draped closely around my shoulder for our sessions. The school staff used to comment on the incongruity of our sizes, and often asked me if I wasn't afraid. "Nope," I would reply truthfully. Where was this terrible animal who had been written up in the initial IEP I received? I never met that young man.

Since that incident, I have stopped reading IEPs until I meet the client. As I said, I scan them, and I do talk to the staff about the student's behavior in class to get a general idea of their issues, but I've learned that the relationship kids have had with their previous therapist is not going to determine their relationship with me. Different personalities cause different behaviors, and sometimes it's better to go in without prejudices. When I meet a child for the first time, I asked them straight-forward questions:

1. What brought you here to this school?
2. What problems did you have at your old school?
3. Who do you live with?
4. What problems do you have at home?
5. What's your best subject in school?
6. What do you like to do when not in school?
7. What did you like do to with your previous counselor?
8. Do you have a social worker, probation officer, outside counselor, etc.?
9. Are you on any medication?
10. Do you have any questions for me?

If the kids say they don't know why they are at my school, then I pull out the IEP and we read it together. Generally the kids tell the truth—the kids who say they don't know why they're at an SED school tend to deny responsibility for their actions, or blame others. Most kids speak the lingo and use the language they've been hearing their whole life. Surprisingly, they usually do not have questions for me. I don't think many adults ask them this question.

Most kids come to counseling willingly the first time. When they do, I explain the rules to them—respect me and my stuff. Short and simple. Complicated rules are too hard for these kids to remember. However, there are the kids who when you go to class and introduce yourself, you're met with a hostile, "I don't need no fuckin' counselor!" When that happens, I just lean against the door in a non-threatening way, and reply, "Okay, I hear you. But you know what, you're going to have to come or you'll get on LOP (Loss of Privilege), so what can we do to make this work for you? Do you know how to play Texas Hold 'Em? How about you come and we just play cards. I promise we don't have to talk if you don't want to."

Usually, the thought of learning something they've seen on television is enough incentive to get them through your door. Chances are, once they are in your office, they'll start talking eventually. I had one student who never came with me for the first three months he was assigned to me. Or he would come and sulk in my office, complaining he wanted a reward or to play basketball. Once I taught him Texas Hold 'Em, he would ask, "Are you seeing me today?" whenever he saw me on the grounds. He rarely talked in session voluntarily, but he would at least answer my questions. Later, when his father died, we had an established relationship so that he could talk about how his father's death had impacted him and his family.

Ironically enough, one of my most difficult students at my school in Watts, who was about ten at the time, was very destructive in my office and repeatedly had to be removed by staff. Honestly, I do not remember what happened to him. I believe he got shipped off to the Halls. Fast forward five years later, and I get a new student at my current school, and when I read the IEP, the name rings a bell—sure enough, it was Arnold. When I went to get him, with some trepidation I must admit, he remembered me from the previous school even though we only worked together a few months. He came willingly, and spoke honestly while we were together. What had changed between now and the previous time I could not say. Perhaps the client just needed time to mature. I wish I could say things worked out, but after three weeks he disappeared, just like he had at the previous school. The point is, the first impression you make with a client is important, because there are only so many "restricted" environments for these kids to go to, and you often see the same clients again and again. The places may change, but the faces stay the same.

When I took a job at a group home, I was responsible for providing individual therapy and group therapy to six adolescent males between the ages of twelve through eighteen. The first week I met with the boys, they quickly realized in their individual meetings that I wasn't like their previous therapists. I played Black Jack, Poker, Rummy, and had other games that they didn't know. However, during group when it was six against one, they decided to test me. I do not remember what my topic was for that first group meeting, but I do remember how that meeting went.

Suddenly, the fairly well-behaved six young men I had met individually turned into a mob of foul-mouthed, sexually inappropriate hoodlums. Every time I opened my mouth to speak, one of them would cut me off, shouting some obscenity while the others giggled uncontrollably. Nearby, a child-care worker stood, listening in. I sat back in my chair observing the boys for a few minutes, and then I shouted at them, "Shut the fuck up!" They stared at me with eyes wide. I went on, "Okay, you guys can cut the bullshit and cooperate, or I can make your lives a living hell. Now what's it going to be?" Several of the boys started to clap, and one of them said, "*All right*! Finally we got a cool therapist!" Clearly, I had passed the test. That does not mean that they came to group every week with a song in their heart, but they at least stopped with the harassment. Or, I should say, they stopped harassing *me* at least. I tried to make group as painless as possible, and they tried to cooperate to the best of their abilities, but after living in a group home for year on end and attending literally hundreds of groups—most of them just didn't care. The best I could hope for was a modicum of respect, and if speaking their language—profanity—was how I had to earn it, so be it. At least I got their attention.

I was to later learn from a staff member that the previous therapist, who had quit after less than two months, did so during group—and had apparently fled in tears saying something to the effect of "fuck this shit." She had used the right words, but at the wrong time.

I actually started working with adolescents in school by chance. It was not something I set out to do; in fact, it wasn't something I knew was possible. I was a trainee, and needed extra hours to graduate, and my clinical director of the agency where I worked asked if I would be willing to go into a high school to work with teens. I'm naturally the kind of person who says "yes" to most anything, just because I figure until I've tried it, how can I know if I dislike it. The school I was sent to was a continuing education high school in the Valley, and I was to work there over a year with teens 14–19. It was, despite the fact that I received no pay, the best experience I ever had.

The difference between the population at a Continuing Education facility and an SED school is that these kids are not required to get counseling. In fact, since most Continuing Education schools have no counselors, counseling is in high demand. It also means that going to see a counselor has a stigma attached to it, whereas, at an SED school counseling is a given—*everyone* has a counselor. However, most Continuing Ed schools have very strict rules, and this particular one was no exception. They had a zero-tolerance policy, meaning that any act of physical aggression, drugs on campus, or truancy and you were kicked out. Kids who go to Continuing Education are often those who miss too many days of school at public school due to truancy, or pregnancy, or are on probation, but their behavior is not deemed severe enough to put them in a non-public school environment.

For whatever reason, I seemed to have an instinctive approach with the kids that I worked with. I would work with a gang member on probation, a teenage mother who was a victim of domestic violence, a girl suffering from anger issues (her arm was broken from punching a cement wall), and other various young people with equally distressing stories.

My approach when the principal of the school would bring a student to me for counseling and we were introduced was along the lines of the following. Usually, because of the stigma attached to seeing a counselor, the student would say they weren't interested. I would respond, "I hear you. But how about this. Let's meet today, just for half an hour, and if you don't like it, you don't have to meet with me again. It's entirely up to you, I promise. If you decide it's not for you, you won't get in any trouble."

I had a 100 percent retention rate with every child who was referred to me. Every student agreed to meet with me the first time, and every child stayed on my caseload until I left the school or they left. When I left the school for a paying job, terminating with my six teens I had worked with—four of which I had worked with for over a year, was the hardest thing I ever had to do.

I made myself as non-threatening as possible in the first session. We met in a makeshift room that was used for who knows what. There was a circular table and chairs, and I often leaned back in my chair with my feet up on the table—hardly the stereotype of a therapist. With the gang banger I used profanity. I told him his situation sucked, and I spoke his language. Of note, while he almost always missed school on Mondays, he was always there on Tuesday—the day I worked at the school. These were kids who were not in the system (with the exception of the banger who had a probation officer he rarely saw), and they had no one else to talk to. Even the teenage mother had no outside services. The kids would talk non-stop for their session, and be shocked when their time was up. Many asked if I could come twice a week. With these students, my approach was predominantly non-confrontational. They needed someone to talk to, and I would listen.

These kids often feel as if they have no control over their lives. Adults are always telling them what to do. One nineteen year-old I worked with had a father who was an alcoholic and a gambler, and spent all of his wife's earnings at the casinos. My client had five siblings she had to get ready for school in the morning, because her mother was at work. My client had become the parent. Unfortunately, due to her age, I could not report child abuse. She was falling asleep in class because she was up at 4:00AM dressing her brothers and sisters. She had nightmares at night, and anxiety attacks during the day.

The most resistant student I met with had served jail time for assault and battery against a teacher. She had been sexually molested, and had been through therapy her whole life and made it clear she did *not* want to see me. Her body language said she would rather punch me in the face. When I said, "Just give me a half-hour of your time—if at any time during that thirty minutes you want to walk out the door and go back to class, I'll respect your decision."

By my putting the seat of control onto these kids and telling them "you can come with me or not—the choice is yours and I will respect your choice," they feel empowered. They do not perceive me as one more adult telling them what to do. As a result, our relationship is built on mutual trust and respect, not resentment.

6

Does Race Matter?

I have been working with hard to reach adolescents for nearly a decade, and I have worked with blacks, Filipinos, Armenians, whites, Hispanics, Tongans, Koreans, and Chinese. Many of them have been of mixed races. I've come to believe that race does not matter—but personalities do. If there is a personality clash, and at some point there will be, no therapeutic relationship will be established. This isn't the fault of the client or the therapist. It's simply a fact of life. Not every therapist is going to get along with every client they are assigned; and if they say they do, they are oblivious to their own countertransference issues.

With this population, where the kids are so desperate to have someone in their corner—someone to depend on who "has their back," skin color is the last thing on their mind. The best case I can use to illustrate this point is Jaime.

Jaime was of mixed ethnicity and I worked with him in Watts. He was maybe fourteen or fifteen at the time, and had been with his previous therapist, a Hispanic, for two years. When I went to get him for his first session, he came willingly, and answered all of my questions politely, but never gave me good eye contact. His body language was not hostile—just distant. I asked my usual, "Do you have any questions for me?"

"Yeah. Are you a racist?" He demanded, making eye contact for the first time.

"No," I answered, surprised by the question. "I tend to judge individuals based on their own actions, not based on a group of people as a whole. If somebody pisses me off, and they're Asian, I'm going to be upset with that person, but I won't become racist against Asians. Why do you ask?"

"Well, I'm a racist," Jaime declared hotly. "I don't like anyone who isn't black or Hispanic."

I nodded sympathetically. "Well, I'm sorry to hear that. But since all the therapists here are white, I'm afraid you're stuck with me, so we're just going to have to work through this."

I took Jaime back to class, and as I walked away I could hear him loudly proclaim in class that he hated his new "white, bitch counselor." From that day on, I heard from my supervisor that whenever she went into Jaime's classroom, he was ranting loudly about his "white, bitch counselor." However, in session with me he was always polite, if distant, and he never used profanity. But when I dropped him off, he made sure I could hear him shout his epithet through the open classroom door (the doors of the classes were often left open for fresh air since the windows did not open).

After working together for approximately six weeks, Jaime brought something up in counseling, the content of which eludes me. I responded to whatever he said, by reminding him of his goal to become a teacher. Jaime stared at me, angry. "Who told you I want to be a teacher?" he demanded.

"You did," I gently reminded him. "When we first met I asked you what you intended to do upon graduating, and you said you wanted to teach middle school. Preferably math."

Jaime didn't say anything significant during the rest of our session. However, at the end of the day when I left the campus, I walked by Jaime's class. The door was open as usual. I heard a voice sing out, "Bye, Jillian! Have a good day!" I turned in surprise to see Jaime grinning, and waving at me. The next morning, Jaime was sitting with the rest of the students where the buses dropped them off, and he greeted me with, "Hi, Jillian!"

In the end, it was not my skin color that mattered to Jaime. It was that I had listened to what he said, and could remember it, virtually verbatim, six weeks later. To an adolescent, who often feels unheard, or that their thoughts or feelings have no value, this was more important than if I had been a long-lost family member. Jaime would end up being one of my hardest-working students, openly discussing his feelings about his family and requesting to see me whenever he was going through a particularly difficult time.

I have found the most effective way of dealing with racial issues, or rather what could be perceived as racial differences from the viewpoint of my client, is to address the elephant in the room.

For example, when I'm working with gang bangers, and they are telling me about something that I, as a white middle-class female have never experienced, I am honest. I'll say, "You know, I can't possibly know what it's like to be going through what you're going through right now. Tell me how I can help you?" Usually, they will say, "There's nothing you can do. Just talking about it helps."

I've had many clients at my school in San Fernando Valley who were receiving additional therapy services from outside agencies quit their services because their therapists would talk down to them. The worst thing to say to someone is "I know how you feel." Whenever one of my students tells me they have terminated services with another clinician I ask them why. Invariably they say, "That bitch kept telling me she knew how I felt. What the fuck does she know about how I feel? She's some white bitch who doesn't know shit about living in the hood." Yet, I'm white too.

I try to find the most common denominator, and connect on that level. If I have a student who's a Hispanic gang-banger, who has just lost another friend in a gang shooting, I may say something like, "I can't possibly know what it's like to constantly live in fear that another person you love is going to die. But I remember when my cousin committed suicide. So I know what it's like to lose someone you love unexpectedly. Your pain will eventually go away. Then when you least expect it, it will come back again. Especially around that person's birthday, or the holidays. Maybe we can find some way for you to honor your friend's memory." I try to de-emphasize our differences, and emphasize our similarities. And I avoid empty platitudes—these kids may come from the hood, broken homes, or abused families—but that does not mean they can't recognize when they are being talked down to.

The kids I worked with at the group home had to see a psychologist every two weeks in addition to seeing me weekly. To put it bluntly—they hated the psychologist. They said she tried to analyze them, talk about their past, and she used psychobabble. With me, they usually spent their full session, saying at least I didn't bullshit them, and I was "real." The psychologist was white, just like me. I never had a conversation with her since our days never coincided, but based on the staff's report, she was trying too hard to go by the book. Kids in residential placement definitely require thinking outside the box if you want their cooperation.

To be quite honest, where I saw the most problem with race was with the parents and the staff in Watts. When I called parents on the phone, or met with them in IEPs, the parents would be prejudiced against me assuming I was prejudiced against them from the start. So while their child would have a great relationship with me, the parents would have a negative attitude before meeting me. I remember calling a father to discuss his son, and he just made a sarcastic comment about how he kept his son tied up to a bed, never fed him, and beat him with chains. "That's what us black people do, right?" I sat there open-mouthed, not knowing how to reply to such an inappropriate response from a parent.

In another instance, I had a child who after undergoing surgery was currently on massive amounts of medication, and had become suicidal. However, when questioned, the mother refused to tell me what his surgery had been or what medication her son was on, saying it was "none of my business." I tried to get the school to get the information, but they refused, saying "blacks don't like whites meddling in their business." I could not get the school staff to understand that if there was a medical basis for the child's sudden suicidal ideation, he needed to be referred back to his treating physician. The staff, however, was more concerned with keeping the peace with the parents, than protecting the safety of the child.

Perhaps the most frightening thing was the racial prejudices that the children were being taught in the classroom at Watts. I'm not certain where the teachers working at the school were educated, but they had their own slant on history that was interesting, if not incorrect. I was in a class one day with a student who was on LOP, and heard the teacher tell the class that "all white people in America come from families who used to own slaves." Well, the little black boy sitting next to me that I was observing immediately swiveled his head around to stare at me in horror. Apparently the teacher had forgotten that many whites in the south didn't believe in slavery and helped the slaves escape to the north. After the abolishment of slavery, there was a massive immigration of whites who came to America after World War I to make a better life for themselves, and during World War II many whites immigrated to avoid persecution.

I have to point out, in defense of the teachers at Watts, that some of the teachers I encountered at my graduate school were just as poorly educated. In a class on Cultural Sensitivity, a black teacher stated,

"All wars are caused by white people invading countries of color, and they always use black troops as their invading force."

Clearly this instructor had not had a world history class in a long time. White man did not invent war—humanity invented war and it dates back to the Egyptian times. Whites have fought whites (just ask any English or French person today where animosity against each other still reigns). Ireland still has conflict against their own for religious reasons. China and Japan have had centuries old conflict with each other, and the territories that they have fought for. How about North Korea versus South? North Vietnam versus South. Perhaps the teacher was talking about how the Americans went into those countries—but we did not start those wars. I think this was a teacher who had lost a relative in a war and was angry about it. She needed to deal with her issue in some place other than a classroom.

I do not want to insinuate that all the black teachers I have encountered are bad teachers. At my current school, the opposite is the case, and four of the best teachers we had were black (they all left for better teaching positions at other schools once they were certified). One teacher in particular stands out, because she was older and diminutive in stature. She had a reputation as being strict, and insisted on all the students calling her Miss Jones instead of calling her by her first name. When the students heard they would be in her class for summer school or fall, they would moan and complain— "But she's so mean!" Inevitably, after a month or two with her, she had their respect and she was their favorite teacher. While other teachers would waste the student's time by showing movies in class or taking their students on field trips to the park, Miss Jones was teaching her students to write business letters and cover letters for their resumes. But what stood out the most was that every time you walked in her class you could hear a pin drop. The students were on task, quiet, engaged, and her class was never loud or out of control. When Miss Jones needed to talk to a student, she took them in the hallway and talked out there, respecting the student's confidentiality, rather than shaming them in front of their peers. When she left the school, many of the students were upset.

The problem I had with the class at my graduate school was that it was a program designed for people who were going to work with children. If this is how they teach cultural sensitivity in a graduate level university, what will happen when these employees go out into the real world with these kids? Will the black teachers be thinking, "Your great-grandparents owned my family?" Will the white aides look at a black child and think, "Your family's ignorant because you get a half-assed education?" The racial discrimination and prejudices need to stop with the educated. If teachers with Masters Degrees continue to spread racism, what hope can we have for a ten year-old gang banger who already hates his life and does not care what happens to himself or anyone else?

The population I work with are already looking for an excuse to get into a fight with their peers. They come to school angry over something that happened at home, and they take it out on the first person that annoys them. I have witnessed little six or seven year-old kids standing in line to go from one class to another, and a student will bend down to pick up a dropped item, brushing their elbow accidentally against the student in front of them. The "victim" will then lash out, screaming, "He pushed me!" even if the other kid apologized for bumping into them. These children are literally volcanoes waiting to erupt, and they do not need teachers or aides fueling the fires with their own racist viewpoints. Educators are in school to teach math, reading, spelling, English, and science—and to correct behavior—not to instigate further behaviors.

I have had a lot of my students ask me "what are you?" I always ask for clarification, because I want to teach them that it's okay to say the "race" word. I remember one little boy, about ten who asked me, "Are you an American?" I didn't know how to take that one, so I answered, "Well, my mother's side of the family is Irish, Polish, and French. My dad's side is Dutch, English, and French. And I was born in Japan. So what do you think that makes me?" He just stared with his mouth agape.

Ironically enough, many of my Hispanic students think I'm Hispanic. My skin-tone tends to have an olive complexion, and I tan easily in the summer. One year when I came back to my Valley school after the August break, I ran into a good friend of mine, a black teacher, in the lobby, and she grabbed hold of me exclaiming, "Oh my God! I know what you did over the break! You became black!" We had a good laugh over that. However, I think that the Hispanic kids like to think I'm Hispanic because that makes it easier to accept why they are so close to me if I'm one of them.

Regardless, I often get asked what I am, and invariably when I explain my cultural background, I get confused stares. I joke that I'm "European Goulash." Some of the kids think because I was born in Japan I must be Japanese. I ask, "Do I look Japanese?" and many will reply in the affirmative. Go figure.

While at Watts, I worked, rather unsuccessfully, with a young black named Derrick. He was always hostile to me, very boastful, and angry. Nothing I did with him seemed to work. I could not form a connection with him. Our school had tried several times to get a black therapist to work there. I myself had several black friends who I would encourage to apply at the school, and their response was always the same—"I worked my whole life to get out of the hood. There ain't no way I'm going back there!" Finally, our school hired a nice, no-nonsense black woman. Derrick had disappeared for a few months. When he returned, I suggested to my supervisor that she switch my client to the new therapist—maybe he would bond better with a black therapist. When Derrick saw me in the hallway, he asked me when I was going to see him. I explained that I had a full case load, and he was being switched to the new therapist. He asked who it was, and when I explained, his response was, "The black bitch! You're dumping me with the black bitch! Man, that's just cold! Why are you doing me like that?"

Where the issue of race does come into play, on a more subtle level, is occasionally you may encounter a child who for whatever reason disowns their racial group, and tries to identify with a different group. Just as there are children who go through phases—sometimes to the point of having a disorder—where they reject their gender, some children will reject their ethnicity. While I have worked with many children who have been of mixed ethnicity—usually they have one black and one Hispanic parent, it is usually children of a clear ethnic group who reject their ethnic origin. I had a Hispanic student, who had significant mental retardation, who insisted he was white. Yet both his parents spoke only Spanish. When I pointed out to him that he was Hispanic, his response was, "I am?" I suspect there may have been some shame issues, or possibly even some abuse issues, although the latter was never confirmed—but for whatever reason, this boy insisted he was white.

I had another child who was Filipino, who started acting black after he had been at our school for over a year. He would walk like a little banger, do hip-hop, give gang signs, talk ghetto slang, use gang language, and whenever I questioned him, he said he wanted to grow up black. This was another case where his parents did not speak English, and I cannot even imagine what his parents thought when their son came home speaking "gangsta rap."

While these children were not able to articulate their feelings about why they were rejecting their heritage, I suspect on some level they may have been ashamed of having parents who were unable to speak English. When you are learning disabled, and cannot get help from your parents with your homework because your parents themselves are unable to read the homework, in a child's mind, the parents seem less intelligent than the role models around them at school. So the students start to emulate not their parents, but the staff at the school.

In the case of the little Filipino boy, the main aide in his class was a black male. In the case of the Hispanic boy, he had a white teacher and a white counselor. Race only becomes an issue when the student becomes conflicted by their identity, either because they do not know who to identify with because they have a white mother, a Hispanic father, and a black stepfather—or they reject their own race out of embarrassment due to problems with first- versus second-generation immigration issues.

7

Attachment Issues

I remember clearly the first time I went to pick up Andy, a 10-year old ADHD male who had been at the same NPS-school since he was six. Andy, I would later learn, was angry, depressed, and came from a very dysfunctional family who made it clear to him that he was an inconvenience at best. When I introduced myself to Andy, his immediate response was, "And how long before *you* leave to get a job that pays more money?" Later, when I was able to peruse Andy's file, I learned that I was his fourth therapist in six years. It was little surprise to me that Andy was unwilling to trust me. In the end, it would be Andy who left me, returning to public school after we worked together for over four years. But it would be years before he would believe that I would not abandon him, and even until the very end of our work together he needed constant reassurance that I cared.

Children who live in group homes, foster homes, residential placement, or for various reasons get kicked out of public schools and placed into the non-public school system tend to have attachment issues. These are children who were often unable to go through the normal developmental stages, and the instability of their home environment is often recreated by the secondary environments where they are placed to theoretically heal and be nurtured.

For two years I worked in a group home providing therapeutic services to six adolescent boys ranging in age from twelve to eighteen. The company I worked for owned six separate homes. In the two years I worked for this company, there were three different managers of the home where I worked. The group home manager is like the paternal figure that the boys should be able to go to for advice and guidance; but how do you trust that figure when the person constantly leaves? The staff, which consisted of child-care workers, was also constantly being changed, either as a result of conflicts in scheduling or personalities, or the staff would quit.

The attitude of the boys in the home was one of "why should I get close to someone? They're just going to leave. Everyone leaves." The therapist that I had replaced had stayed less than two months before quitting. The group home eventually terminated me, and all the other therapists who provided services at the other five homes in order to outsource the boys to a cheaper facility. Ironically, I was being paid only slightly more than what I made when I was an intern. One of the boys, who was seventeen and had been in the home since he was ten, came out on my last day and stood and watched me drive off, waving as I left. The entire two years I had worked at the home I felt I had made no connection with this particular boy. Perhaps I was wrong.

The situation in schools and residential facilities can be even worse. Many schools hire teachers who are working on their credentials, while residential facilities hire interns who are working on their hours toward licensure. As soon as these professionals get their necessary credentials or licenses, they leave to greener pastures for better pay. Others are simply unprepared for the environment they are getting into and quit within a short period of time.

The school where I currently work is notorious for losing staff. Part of this is a result of low pay. It is also due to the challenging nature of the population. I often joke with my colleagues that our environment is not a school, but a cross between a mental hospital and a jail. My reasoning for this is that the number one concern is the student's safety, followed by behavior modification. If, after all that is said and done, there is time for education—great! Unfortunately, most teachers think they are being hired to teach. As a result, many staff quit after one day—some last a few months. Then there are the ones who are unable to control their classrooms and are fired after one year. The result of all this is that the children learn not to get attached to a staff, because they know that they might not see the person the next day.

Psychologically stable adults tend to take attachment for granted. We wake up each morning secure in knowing that, barring some extraordinary event, when we go to work the same co-workers that we've had beside us day after day will still be there. When we come home at night our spouse and children will be safely waiting for us. When we attend classes our classmates and teachers will be present.

Imagine going through graduate school and every week a different teacher showed up, and you didn't know who would be waiting for you when you went home. How would this impact your anxiety level? How well could you concentrate on your class assignments? How would this disrupt your sleep? How would it impact your overall health?

Now try to imagine that this is all you've known since you were a child—and you are still a child and are expected to cope with this. You live in a group home and all your possessions have to fit into a duffle bag. At a moment's notice (and most kids in a placement get no warning in order to avoid physical altercations) you can be moved to a new location. You do not get to say goodbye to your housemates, your schoolmates, and you now have to live with strangers and maybe transfer to a new school. This is your world, and it will be until you are deemed by the state that you are an adult and capable of being responsible for yourself.

Let's take this one step further. Imagine all the above, but add to the scenario the following. You are mentally retarded or have psychotic features. Or are autistic. And this is the question running through your head—"Why don't my parents want me?"

Over half of the children I work with live in some type of foster placement. This may include children who live with extended family members (aunts, grandparents, or elder siblings). These children often struggle with depression due to their feelings of abandonment by their biological parents. Many of these children have never known their biological parents, or the last time they saw them was when they were a toddler. Their surrogate parents often make it clear to them that they are a "burden," "difficult," or "unwanted."

One of the children on my case, Ronny, has lived in foster care most of his life. Ronny is one of those sad misfit kids who does not fit in. He has always struggled to make friends, but because of his peculiarities, he tends to be victimized and teased, and as a result suffers from extreme depression and low self-esteem. As if Ronny does not have a hard enough time fitting in with his peers, Ronny also suffers from auditory hallucinations and delusions. Most of the time when Ronny talks he is nonsensical. However, in session, Ronny is able to express himself very well. He is sad about the fact that he hasn't seen his real father since he can remember, and he wonders desperately why his family gave him up. Ronny may suffer from delusions, but he is smart enough to know it has to be because there is something wrong with him. "Maybe if I was smarter," he asks me.

One of the biggest mistakes I ever made as an intern was when I worked with a young boy, Michael, who appeared unresponsive in counseling. Actually, Michael's affect was often blunted, but I had been working with him for over a year, and had made no progress with him. I made the mistake of personalizing his lack of progress, and assuming it had to do with me, and his lack of attachment to me. In counseling, he was angry and guarded. He often gave me responses that appeared to be manipulative, as if trying to please me. I would later learn otherwise. I had spent most of my time in supervision discussing the case, and my supervisor agreed that at the end of the school term, I could pass Michael along to a new counselor starting at the school.

Summer school started, and Michael missed the first two weeks. When he started late in term, I went to him with his new counselor, explaining to him that I was no longer his counselor, and he would be seeing Miss Sara. Imagine my surprise when this stoic eleven year-old, who had never displayed any emotion other than anger in my office burst into tears and threw his scrawny arms around me in distress. I felt as though I had been stabbed in the heart. Unfortunately, because Michael had missed the first two weeks of summer school, my case load was already overbooked, and there was no way to correct the situation and reverse the decision. Eventually, Michael's therapist would leave and he would return to my caseload, but I carried the guilt of that decision with me in the interim. However, it was in hindsight, the best learning experience I could have gone through.

Since that mistake, I have learned to let the clients make the decision to leave me. I have also learned when to honor their decisions and when not to. Unfortunately, there is no clear cut rule for this. Often, the decision is based on instinct. It depends on multiple issues—the child's diagnosis, how long they have been with you, the strength of the therapeutic relationship, what has precipitated the request, and is the child making progress?

I was assigned a twelve year-old girl, Shanaya, who had come to my school with a history of physical aggression, being sexually inappropriate with boys, truancy, and drug usage. Within two weeks of meeting with her, there was an IEP held for an AB3632 (residential) referral. The Department of Mental Health had already evaluated Shanaya, and recommended she be removed from the home due to the lack of structure. Shanaya was smart, and easily able to manipulate her grandmother, her legal guardian. Ultimately, the grandmother backed down and kept Shanaya in the home, as is often the case when kids get out of control. Parents feel too guilty to put their children in the structured environment they need.

Shanaya would go on to continue with her truancy at our school, throw tantrums at school causing her to be physically restrained on a weekly basis, and at home she would destroy property regularly—including windows, laptops, and cell phones. No matter how much she destroyed, her grandmother replaced the items. Shanaya felt academically superior to her classmates and thought she should be allowed to return to public school, and every few months an IEP would be held to determine whether or not she was qualified to leave the NPS system. Whenever Shanaya's grandmother appeared close to removing her granddaughter from her home, Shanaya would turn on the crocodile tears as though on cue. Later in session, she would laugh about how she had her grandmother wrapped around her finger. When I pointed this out to her grandmother in what was the fifth or sixth IEP in two years, Shanaya started screaming at me, "You fucking bitch! You're lying on me! I hate you! I'll kill you! I want another counselor. You can't say that. That's a violation of my confidentiality." Her screams were so loud that people in the nearby offices could hear her from behind the closed door.

Over the course of the two years we worked together, Shanaya was always demanding another counselor. She would storm out of my office swearing I was an "evil bitch," or other such epithets. But the following week she would be gathered with the rest of her classmates in the morning when the kids unloaded from the bus, looking for me and asking, "Miss, are you gonna see me today?" When she had a photo taken with her best friend at the mall, she gave me one for my office.

After this particular IEP I told my boss who was in charge of assigning students to therapists to hold off until the following week before she reassigned Shanaya. Sure enough, the following Monday, there was my favorite borderline hanging her arm around my shoulder asking when I would see her.

"But you said you wanted another counselor," I teased her.

She just laughed, tossing her hair. "Oh, miss. You know I didn't mean that. I was just mad at you cuz of what you said in the IEP."

On the other side of the spectrum are the clients who will attach indiscriminately, within the first five minutes of meeting any adult. These are the children who have no boundaries, and will share their life history to any stranger who gives them any attention. Teachers, therapists, and social workers become interchangeable—and these kids exude a desperate neediness that can be overwhelming. With younger children, this lack of stranger anxiety can be dangerous at its worst. These are the kids who upon meeting adults, or peers, immediately start talking about how they have been sexually molested, have an eating disorder, or have been in a mental hospital for cutting themselves. This is not to be confused with children who are attention seeking; these children are reacting normally to their attachment needs, by attaching indiscriminately. However, as these children get older, if they do not learn to control these self-disclosures, their entrance into adulthood will be difficult as they try to obtain jobs or transition into college.

Why is attachment with adults at school so important? For the at-risk population, the teachers, aides, and counselors that they see at school five days a week may be the only healthy adult role models they come in contact with. Children in public school who have an intact family, or even children in a single-parent family, may belong to after-school programs such as day-care, boy-scouts or girl-scouts. Most SED children do not get to take music, art, martial arts or sports lessons where they can have a healthy adult role model. The children in foster care or placement rarely get to participate in extracurricular activities. The reality is, the majority of the foster families take these children in for the monetary benefits only, and do not have the children's best interest at heart. The same is often true of extended family members who become the child's legal guardians. Often I see children raised in environments that are only slightly less abusive than their family of origin.

Therefore, when they are at school and they have to cope with a new teacher every year, or the aides in the classroom being changed frequently, it becomes virtually impossible for these children to form a healthy attachment to an adult role model. This was the case with young Andy, where I repeatedly suggested to his father that Andy be enrolled in a big-brother program, a softball team, or boy scouts in order to have a male role model that he could attach to given that his father worked long hours and complained he did not have the time for his son. His father never heeded my suggestions, but what was important for me was that I tried. It was also important for me that Andy knew that I tried.

Often when I feel I am not making a connection with a higher-functioning client, I ask them if they would prefer another counselor. I do this in a non-threatening way that lets them know that I am not rejecting *them*, but that they seem to be unwilling to talk to me, and I want to know if they would prefer someone else. Over 80% of my caseload is boys, so I will often ask them if perhaps they would feel more comfortable talking with a male counselor. Invariably the response I get is, "No, it doesn't matter. I wouldn't talk to a guy either."

I have learned over the years that it is not what takes place in the office that is important, but simply the fact that I show up each week and can be relied on to be there. It is that "secure base" John Bowlby referred to that I provide that matters. One reticent male client I worked with in school for three years graduated, and actually passed his high school exit exam. Carlos was very anxious to know that I would be there on graduation day to see him graduate, and the last session we had he talked excitedly about how he was dressing up in a suit, and going to throw his cap in the air.

Graduation day came, and I watched as he received his diploma, clapping when his name was called. Six months later, I received a phone call from the front desk, saying that someone was at the school to see me. I didn't recognize the name. The receptionist said, "Just come down stairs." When I walked down, there was Carlos, with a big grin on his face. He said, "Hi, Miss Jillian, how are you?" For the first year that I worked with Carlos, he could never remember my name, and called me "counselor." He had always explained that my name had too many syllables and was too complicated. That he remembered my name after being gone several months spoke volumes.

Standing outside looking at the schoolyard, Carlos told me about his search for a job, and how he was doing in college. He told me about the grades that he had received for his midterms, and what his concerns were about his finals. He told me about his family, and what his plans were for his future. I gave him a quizzical look and said, "Do you realize you have just said more to me just now than you said in one year of counseling?" He shrugged and said, "I didn't have anything to say before."

8

Tools of The Trade

In graduate school, there are usually classes on Play Therapy, Art Therapy, Music Therapy, or Therapy with Children. Very few universities have courses on working with adolescents or working with mandated clients—that is, clients who are required by law to receive therapy.

When I went to college, I had a Therapy with Children course taught by a well-intentioned clinician, whose office was in Beverly Hills. She showed us the requisite videos from the 1970s of the pioneers of play therapy doing sand tray work with white adolescents who were "troubled" over their parents' impending divorce. We learned in class how to do the standard House-Tree-Person drawing, sand-tray work, puppet play, and free play using dolls representing family members.

It was not until I took a course in child development, and the instructor talked about working with children with Autism that I realized my class on Therapy with Children had been lacking something. What did you do with a child who lacked social skills? Or did not want to, or was unable to, engage with the therapist?

I was, unfortunately, to learn the answers to these questions on the job. One thing I did learn quickly was—keep it simple. In my course on Art Therapy, we learned about collages, Mandalas, string art, and various other methods of creative expression. That was all well and good—and I did use these techniques with adults in private practice, but in lower socio-economic environments—telling a child to draw a House-Tree-Person can be down right overwhelming. Never mind something as esoteric as a Mandala!

Imagine you are going for a job interview, and the job is to work for NASA as a secretary. During the course of your interview, your prospective employer hands you data on a planet's mass, the rate of speed light from the nearest star travels to the planet, and the point of distance between the star and a third planetary body. Then he asks you to perform a complex mathematical equation based on the information you've been given. You would probably feel intimidated and judged. After all, you just wanted to be a secretary.

Since many of the kids I work with come from single parent homes, often are on food stamps or welfare, or their parents are more concerned with spending money on other things such as drugs, these kids may have never owned a box of crayons. Being asked to draw, while being watched by a therapist, is akin to performing a physics calculation during a job interview. If the child *does* feel comfortable enough to do the drawing, he or she will spend half his session looking through the box of 64 crayons trying to find the *red* crayon. Cardinal red, cherry red, magenta, or red-orange won't work. These kids are only comfortable with the eight or ten standard set of colors.

When I worked at Watts, I was at a school for kids between the sixth and twelfth grades. A group of art therapists from the elementary school joined our supervision school, and all of them said the same thing—the kids were unable to do any of the art therapy exercises they presented them with. They had to rearrange their activities to accommodate the client base.

Rather than using complex art supplies such as pastels, paints, or other advanced mediums, I use coloring book pages and stencils. They are economical in that they can be used repeatedly (I photocopy the coloring book pages), and they are non-threatening because there is no pressure on the kids to perform on an artistic level. I do have the elaborate 72 set of colored pencils, but mixed in are three sets of the basic ten colors, which are used the most frequently. My higher-functioning clients use the more "exotic" colors—the lower-functioning clients stick to red, yellow, green, blue, black, brown, orange, and purple. Stencils allow versatility in that the children can recreate the stencil page exactly, or mix and match multiple stencils to create their own "page," depending on their comfort level. Stencils are also great around holidays, because the kids love to make cards for Valentine's day, Mothers and Father's day, Christmas, and birthdays. I had one little boy make a card for his grandfather who was in the hospital after a stroke.

Many of these kids have perfectionist tendencies, and will get frustrated if they are coloring and they go out of the line. We can then discuss frustration tolerance, and why they need to be perfect. If they get angry and give up, we can then talk about anger management or perseverance. Coloring may sound simplistic, but it can stir up a lot of issues—especially if you've never had the opportunity to do it often. Other kids can color and focus the entire session. I'll praise them, saying, "Wow, imagine how good your math would be if you worked as hard on your math as you do on your art work?" One time I was in a class with a student of mine who has amazing concentration when it comes to his artwork, and he was on punishment for failing to do his English. I walked over to him after watching him pout and sulk for twenty minutes, and whispered, "I want you to pretend you're telling a story with your sentences, just like you tell a story with your drawings. I know you are smart enough to do this, because you tell great stories with your pictures, so you can do it with your words if you try as hard as you do in my office." I walked away and when I looked through the window he was sitting up right, biting his lip in concentration the same way he always does when he draws.

Adolescents who have been in Juvenile Hall, or kids who live in foster care or group homes, tend to love games—particularly card games. The one caveat is that before playing any game where there can be confusion about the rules, state how you play the game. I have many kids who have different rules to a game. Mancala, for example, can be played multiple ways. I tell the kids, "Sorry, it's my office, it's my rules." The reason for this is simple—I do not have the time to remember what rules I use with which kid, so every kid has to play by my rules. Therefore it's best to clarify the rules before starting a game. Speed, a popular card game, can be played with or without doubles, and I always clarify that it is without doubles. Rules to these games can be found on the internet, or a bookstore—better yet, have your student teach you—it's yet one more chance to empower them.

Blackjack, poker (either regular or Texas hold 'em), speed, rummy, kings in the corner, thirty-one, and of course war, are popular. I try to avoid games where kids can cheat easily and I can't keep track of them. For example, I do not play go fish or battleship—it's too easy for kids to cheat. It's not that I'm unwilling to deal with the issue in session, it's just that there are so many games available where you do not have to worry about the kids cheating. Uno, Mancala, dominos, chess, Sorry are also popular.

Othello, which is a cross between checkers and chess, is easy to teach and I've found middle-school and high-school kids love it. Backgammon is also popular and most kids aren't familiar with it. Kids generally like being exposed to new games. Yahtzee and scrabble are good for older kids. For the younger ones, or the ones with significant cognitive deficits, there is monopoly junior, candy land, chutes and ladder, connect four, and memory or concentration (matching game).

On a more personal note, I won't have a game in my office that I do not like to play. For example, I hate checkers and Trouble (it literally gives me a headache due to the noise). I figure if I have to play games daily, I'm going to at least play games I can tolerate. I also set limits to buying games. Every kid who gets assigned to you in a school had their favorite game with their old therapist. They want to hang on to this game as a "transitional object" when they get assigned to a new counselor. So as a therapist you continually hear, "Can you buy Such and Such, game? That was my favorite! I loved playing that with my old therapist." I use this opportunity to talk about grief and loss over their old therapist leaving, and I gently explain that I'm sure they'll find different games in my office that they'll like in time. Usually, after three months, the kids stop asking for the game because the therapeutic relationship has been established. If I bought a new game every time I inherited a new client, I would be broke.

On a similar note, there are certain craft items I just won't use. When I was inexperienced I spent the money. Now I realize I do not need these expensive items. PlayDoh, Crayola Modeling Magic, plastic beads for jewelry—all of these types of items that can only be used once, and then the kids want to take their creation home with them, can be expensive—and messy. If you want to invest in them once a year, say at Christmas, or during summer school, as a treat, you can. But I can guarantee the kids won't understand or appreciate it. Instead, they'll whine, "But you had it before! Why won't you buy more?" Because these kids can be excessively needy, and many of their parents teach them that love is bought through material items, rather than *quality time*, I try not to recreate that environment in my office.

In terms of toys, I have duplos—I lucked out and purchased several large sets of duplos with Winnie-the-Pooh characters, which many of the little kids love building complex landscapes with. I also have sand-tray animals, without the sand tray. I have over 150 animals, often with whole families of different breeds (lions, deer, horses). This allows for unstructured play that can become representative of aggressive or nurturing tendencies. I have some playschool cars with characters (police, fireman, workman), and also matchbox cars. Kids can be very creative. I have the game Jenga, but most of the kids like to use the game pieces as building blocks. In a similar fashion, dominoes are often used to make domino trains that they then knock over.

I have, over the years, been able to acquire many games and activities that are good assessment tools. I stumbled across a stacking game called Chairs (sort of the opposite of Jenga), which is excellent for assessing ADHD. The box contains two dozen chairs of different shapes and weights, and I have the kids stack as many as they can without them falling. As with the coloring, some kids will get frustrated after one attempt and quit; others will spend the entire session trying to stack them—analyzing the problem when they fall, and trying to correct it. Jacks can be great for assessing hand to eye coordination, without leaving the office for the playground. Little kids, especially boys, love them. Most have never been exposed to this game, and will usually try it repeatedly. Magnetics is also popular with boys, and is a good assessment tool for frustration tolerance, problem solving skills, and spatial perception skills.

Catch-the-Match is another excellent assessment tool. It's a collection of 15 cards with matching images on each card, but the images are in different colors of two—you have to find the one item (hat, pencil, boat) that matches on both cards with the two colors in the same place. It requires total concentration. Interestingly, I had one student, a young man with mental retardation, who was unable to play most games—but he had excellent visual skills, and he could play this and win about half the time. He was also able to win at concentration (a matching game) about half the time. So although he did not have the cognitive ability to grasp rules to complex games, anything that was visual, he was strong at.

Fractiles are another excellent art tool. They are magnetic pieces in different colors and geometric shapes that come with a steel board (if you get the smaller set you can use a cookie cutter sheet) that the pieces adhere to. There are instructions for making different objects such as pinwheels, butterflies, or spaceships with the pieces, but the trick is putting the pieces in the right direction. Fractiles can be found at most museum stores, learning stores, or on the internet (www.fractiles.com). One of the advantages of using toys such as Fractiles and Chairs, is that when not using them as competitive play, it teaches children to ask for help when overwhelmed, a common problem in the classroom due to shame issues.

Puzzles are also an excellent assessment tool. I had a student in middle school who was academically on track, but was unable to build even a 24-piece puzzle without becoming frustrated. If I brought out a puzzle, he would literally run screaming from my office, yelling, "No puzzles, no puzzles!" George, a high-functioning Asperger's child, had very poor spatial perception skills. He had never played with creative toys as a child, and had spent his entire life on video games. One year, when George had a melt-down after losing at a board game he jumped on the box, and I told him that during summer school as a consequence he could not play games at all. He threw a fit, and seemed at a loss as to what was left for him to do in session. "There is art work, toys, puzzles." He spent his summer struggling to play with toys—imaginative play was simply outside of his realm.

As I mentioned earlier, letting kids teach you games can empower them. I had a chess set for years, and it was sitting at my school at Watts. While I understood the basics, I had never really been able to play. One of my kids there actually taught me the game (he succeeded where my boyfriend had failed!). He was able to explain why my moves were good or bad, and made an excellent, patient teacher. While I am still by no means a strong player, I'm good enough to play with the children that I work with and my skills have served me well. More importantly, I let the student who taught me know every time we played what a good teacher he was. I praised his strengths—his patience, the ability to articulate himself, and that he did not get frustrated and angry playing with someone obviously inferior to him.

Another student I worked with at Watts was an excellent Scrabble player. I have always been a strong Scrabble player, but even Brian taught me tricks I never knew, and we spent our entire time together battling out our vocabulary skills on the scrabble board every week until he returned to public school. As a side note, for kids who are intimidated by Scrabble, but want to work on their vocabulary and reading, Boggle is a good alternative. Word search books are also good. Like coloring books, you can photocopy the pages and have them on hand, and then talk while the kids do the pages.

At my school in the San Fernando Valley, I have a white board in my office. It is situated against the wall where the students sit, right next to my desk. When I originally moved into the office, the white board was across the room, and the kids rarely used it; once I moved it to where they sit, they started using it every week. A white board is less threatening than paper for art work. You can erase things more easily, and the kids can draw while also playing a game. The gang bangers tag on it (any inappropriate drawings have to be erased before they leave—by them). But often kids will spend their whole sessions drawing elaborate pictures, begging me to keep it up there. I always make the same reply, "It will stay up there until some other student erases it—but I promise *I* won't erase it."

The white-board is often excellent for games like tic-tac-toe or hang man, without wasting paper. One particularly precocious little girl, Coryn, used the board to create her own games. It started with her drawing patterns, and then leaving blanks and asking me what patterns would come next in the blanks (it was actually the same sort of sequencing used in standardized testing). She would then ask me to create one. Then I created a game of "which one doesn't belong," where I would write a group of four words and she would have to circle the one that didn't belong. For example, lion, elephant, dolphin, monkey (dolphin not belonging because it lives in the water). We continued to create different types of games using the board, and every few months Coryn finds a new way to use the board.

Several of my older students like to use the board to practice their penmanship. Most of these children are academically behind their peers in public school. So you can have teens in high school who do not know how to write in cursive. They will practice on the board, and then I can use another color marker over their words to correct their letters.

I also have several children's books in my office. Some of the kids simply enjoy reading aloud. Others like to work on their reading in the privacy of my office where they won't be embarrassed by their peers listening to them. Some of the best books that kids enjoy and aren't too obvious in their message include *Today I Feel Silly and Other Moods that Make my Day* by Jamie Lee Curtis, *The Pain and the Great One* by Judy Blume (which deals with sibling rivalry), *If You Give a Mouse a Cookie* by Laura Joffe Numeroff, *The Missing Piece* by Shel Silverstein, *Where the Wild Things Are* by Maurice Sendak, *Tacky the Penguin* and *Hooway for Wodney Wat* both by Helen Lester (dealing with being different), *There's an Alligator under my Bed* and *There's a Nightmare in My Closet* both by Mercer Mayer (deals with problem solving and facing fears), *Alexander and the Terrible, Horrible, No Good, Very Bad Day* by Judith Viorst, and *Dr. Seuss' Oh, The Places You'll Go!* (motivational). While I spent a lot of money on books on abuse, divorce, grief and loss, bullying, and other more heavy, psychological matters—the kids never picked those out to read.

Most of these kids come from families where they were never read to as a child, which is why many of them get so frustrated or stubborn about reading. "It's boring. It's for sissies." But I've found that if I offer to read to them, they'll jump at the chance. So for the kids who are seriously delayed in their reading skills, I'll read an entire book to them, making the experience as enjoyable as possible. Then the next time they ask for me to read, I'll agree to read, but only if they try to read every other page. Over time, the ones who are really motivated to return to public school or get their diploma, will start reading entire books on their own, without being asked.

9

Setting Boundaries

John was assigned to me when his previous therapist had left the school. John was approximately seventeen when we started working together, and he had moderate mental retardation. Whenever I'm assigned a client who came from another therapist, I always ask, "What did you enjoy doing with your previous therapist?" It allows me to gauge what type of session structure the previous therapist set, and also what the client likes to do. John replied that his previous therapist, a male, took him out to eat lunch every week. While this was unusual, there are many therapists who use the "contract" system of rewarding clients by taking them to buy something at the 7-11 or 99 Cent Store if they maintain a certain behavior for a designated period of time. However, this was the first time I had heard of a therapist taking a client to eat lunch regularly.

I explained to John that I did not do contracts, and that I did not give student rewards for good behavior in my office. I made it clear that I expected good behavior, and misbehavior would get him sent back to class. In first sessions with kids, I tell them, "I have only one rule—you can express yourself anyway you want, including using profanity (this latter part is only for older kids). However, I expect you to respect me, and to respect my things." While that sounds simple, it actually covers a lot of ground. Week after week John would ask me for food. I explained I had none in my office, which was true. At one point, John crossed over my body to get into my desk drawer and I stopped him.

"What are you doing?"

"I want to see what's in your desk," he replied.

I told him to sit down immediately, and I explained that he was allowed to sit on his side of my desk, but he was never to cross over to my side of the desk. I asked him if he liked it when people went through his personal possessions, and went on to explain that he was to never go into a teacher or staff's personal possessions, because it was rude and disrespectful.

John's teacher told me that John was often sexually inappropriate with female staff and peers, and it was not long before this behavior appeared in session. First John kept moving his chair from his side of the desk next to mine, so that our seats were touching. Every time he did this, I reminded him that he was to stay on his side of the desk. He would get exasperated, and ask why. I would explain that it was a rule, and he was required to follow rules. Then on one occasion, he stood up and dropped a game piece when I was leaning forward to make a move down my shirt, and then tried to retrieve it. When you work at a school with hair-pullers and autistic children, you develop lightening reflexes and I grabbed his wrist, telling him he was being inappropriate. I took John back to class, the whole time he argued that he had not done anything wrong and that I was being unfair. I made his teacher aware of his actions, and he was promptly put on LOP (loss of privilege). I made certain to document everything in my notes when I returned to my office.

As John tried harder and harder to manipulate me, he became more and more frustrated with the structure of his sessions. When he turned eighteen, he complained that he was a man, and he wanted to spend the session talking—but he wanted me to do the talking. He often asked me inappropriate questions of a personal nature, and I kept reminding him that he was in session to talk about himself. Then in the next breath John would complain that I didn't take him out and buy him candy. We had many long talks about his conflicted feelings about wanting to be treated like an adult, without any of the responsibilities, while still wanting the advantages that went with being a child.

Meanwhile it was clear that John did not see our sessions as therapy, but he wanted to turn them into dates or at the very least, outings. The more structure and rules I set, the more petulant he became. During this time, his previous male therapist had returned to the school, and John would frequently whine, "Maybe I'll go back to Mark. I had more fun with him." I made it clear to John that if he wanted to return to his previous counselor, he was free to do so, but he would have to make the request in his IEP and in a respectful, mature manner. Approximately two years after we had worked together, he did. John lasted approximately three months or so with his prior therapist before dropping out of school all together.

In private practice, clients come to an office—a neutral environment—to see their therapist for an hour, then go they home. Sometimes there is even a separate entrance and exit to the office, so the clients do not even know who the other clients are—they have no chance of meeting each other in the waiting room. Or there is a ten minute interval between clients and clients rarely know which therapist a person in the waiting room is there to see. The client sees the therapist in an artificial setting once a week, and then goes home to their life.

When working in a school, group home, or residential setting, those normal boundaries and false environments created by the private practice office do not exist. For example, at my school, all the kids know who everyone's therapist is. They discuss their counselors in class and on the bus. When I walk into a class, I may have three or four children in that class on my caseload. Obviously, client confidentiality in terms of who my clients are does not exist.

Furthermore, the clients see me in my work environment. They see how I behave with the staff and my friends at work. And horrors of horrors, they see how I behave with the other clients on my caseload when we're walking to and from class. As one student said to me, "You don't act the same way you do with me as you do with other kids you see." I replied, "Do you act the same with all the kids in your class?" We went on to have a discussion about how different personalities bring out different aspects of people.

In the case of John, because the previous therapist had set no boundaries with this student, and had led John to believe that therapy was all about having fun, and going on outings, no amount of structure in the world could allow me to undo the damage the prior therapist had done. The therapist had no boundaries with the student, and the student in turn had no boundaries with the staff or his peers. Perhaps John had no boundaries to begin with—but if his previous therapist set some limits when John was younger, perhaps John could have benefited from the therapy process.

Different facilities have different rules about taking students off-campus. I've worked at facilities where there have been strict, no off-campus rules, to facilities with very lenient rules. However, I figure that these students get enough lack of structure at home, and I don't need to add to that sense of confusion. It is possible to keep therapy fun, and keep it confined within four walls.

Because of the lack of physical structure in an open setting, setting emotional boundaries with the clients becomes even more important than when working in a regular private practice. When working with emotionally-disturbed children, setting boundaries with clients is similar to parenting: Keep the rules simple, be consistent, and let the staff know your rules. If the rules are simple, the kids can remember them. They can't double talk you, and manipulate you into thinking you said something else. Challenged kids are like toddlers— they know that if they tantrum long enough you'll give in—once they do, they own you. Then they will tantrum every time. "But last time you let me!"

Convey the rules to the other staff, so the kids can't "good parent, bad parent" you. You'll tell the child they can't do some activity, because their behavior in counseling was bad, but they'll give a sad tale of woe to a staff member they know they can manipulate, and next thing you know they've gone behind your back and gotten what they wanted.

Boundaries, or limit-setting, is imperative to survive this population without burning out. In a residential facility or school environment, the children see you multiple days—four to five if you are there full time. When I walk across the grounds of my school I continually hear, "Miss Jillian!" or "Hey counselor!" The little kids will come up and wrap their arms around my legs, looking up with puppy-dog eyes, begging, "When are you going to see me?" knowing full well I saw them the day before and they don't get to see me for another week.

While you do not have to cut off your emotions and become a robot, you do need to set limits. Let your kids know that when you are with another child, transporting him or her to class—it is *that* client's time. I had a borderline who would run up to me to hug and talk to me whenever I was with a client. I had to tell her, "You're time with me is when we are in my office. No one else interrupts our time together, I expect you to respect my other clients' time with me the same way they respect my time with you." I tell kids they can wave and do shout outs, but all they will get in return if I am with another child is a wave—no words. Half the time I walk around campus feeling foolish because I'm walking forward, waving over my shoulder to someone I can't see. It's a difficult balance to respect every client's feelings.

After eight years working in the SED school setting, where I have seen numerous therapists come and go—some after a day, some after a few months, I have seen many examples of how therapists fail to set appropriate boundaries. The problem is, not only do therapists often set inappropriate boundaries with their own students, often they set inappropriate boundaries with the staff or other students, and thus come across as unprofessional.

Children in an SED setting can sniff out an inexperienced or easily manipulated therapist like a bloodhound can track out a fox. These children are often borderline, or at least have borderline tendencies. Whenever a child comes to me and says, "I want *you* to be my therapist," I reply firmly but politely, "Well, that's very sweet of you to say so. But if you want to change therapists, I suggest you tell your therapist why you are unhappy with them and try to resolve your differences." Needless to say, these follow-up conversations never take place. There is a saying amongst therapists—the grass is always greener. It is easy for a student to look at another person's therapist—the one who isn't confronting them on their behaviors, and think, "Wow, that therapist seems more fun than *my* therapist." Students will do the same thing with teachers—they love their teachers, until the teacher tries to correct their behavior or the "honeymoon" is over. Suddenly these students demand a change in classroom.

Unfortunately, inexperienced therapists are often flattered, and strike up a relationship with the student "on the side." One therapist at my school, Linda, kept coming to me repeatedly saying, "Oh, so and so, on your caseload is so interesting. We had the most fascinating conversation." Linda was working at the school full-time, and theoretically should not have had any spare time to have conversations with my students. Linda was working with the younger sister of my client, Maria. One day Linda came to me telling me she had a long conversation with Maria, and afterwards Maria was upset and crying and Linda was concerned and thought I should know.

I confronted Linda on her behavior, asking her why she was talking to my client without consulting with me first. She tap-danced around the issue. When I asked her what she had said to Maria, it turns out that her conversation had been in direct conflict with Maria's counseling goal. I pointed out to Linda how her behavior was inappropriate, and her conversation with Maria could have sabotaged the work I was doing by telling my client to do something contradictory to her IEP goal. Linda was apologetic.

I continued to see Linda having conversations with my students, and I had to confront her about it repeatedly. Eventually, it got to the point where I had to spell out to Linda the damage she was doing. I knew that Linda was looking for another job, and was planning to leave the school. I reminded her that these students already had attachment issues, and by her forming relationships with them when they already were receiving counseling was inappropriate, especially given that she was planning on leaving. Was Linda going to have a termination session with all of my students in addition to her own? I pointed out that she did not have access to the students' IEP files and did not know their counseling goals, and could be giving them conflicting information from what we were discussing in their sessions, thereby sabotaging my work with them. I asked her if she was consulting with their parents about their behavior? Their teachers? Would she be responsible and file a child abuse report if they confided abuse to her? No? Then she needed to keep her boundaries, and worry about the kids on her own caseload, and let the other therapists at the school worry about their respective caseloads.

It is very easy to get caught up in the laid back atmosphere of the school and feel that all the kids there are your "friends." To somehow slip back into those carefree college days and just "hang out." But never for a minute forget that these are troubled kids who need mature, responsible adult role models. They do not need adults who blur the boundaries and act like friends. Most of them already have parents who do that.

I remember when I was an intern, and a friend I graduated with was interning at another school. Jeri and I were out shopping at the mall and she said, "Oh, I have to buy a present for one of my kids—it's his birthday." I stopped dead in my tracks. "You buy your clients birthday presents?" It turned out, Jeri also bought them Christmas presents. She was surprised that I didn't do the same for the kids at my school. For one, she only had fourteen kids on her caseload, while I had thirty. She was salaried, while I was hourly. But aside from the monetary issue, it just did not feel right. I couldn't let the matter go. Driving home, I asked her why she did it.

"So they'll like me," she replied.

Huh. That stunned me. My first thought was that my friend needed therapy to work on some major issues. My second thought was that you can't buy love. What I said out loud, however, was, "If you were in private practice working with kids, would you buy them presents?"

Jeri thought about it and after awhile replied, "No."

"Then why do you feel it's okay to do because you work at a school?" I asked.

"That's a good question."

That conversation has served as a good rule of thumb in my career—if you wouldn't do it in private practice, do not do it in another treatment setting. Obviously there are going to be some exceptions, but you can usually tell what they are by asking these questions—does this serve a therapeutic purpose, and is it in the best interest of the client? If the answer to both parts of the question is yes, you are probably okay. I seriously doubt that Linda, if in a private practice, would have sidled up to the clients in a waiting room, and struck up conversations with them. Stealing clients from your colleagues in private practice is a sure-fire way to get kicked out of an office—fast!

No doubt my friend Jeri, without even realizing it, was recreating in her relationships with her clients, the same relationships these kids had with their parents who probably tried to buy their affection with gifts. I've seen many parents who make up for lack of quality time with their child by buying elaborate gifts—cell-phones, I-Pods, laptops, expensive clothes. This is especially the case when the parents are recently divorced, or it is a single parent attempting to make up for the fact that the other parent is dead or in jail and they are trying to compensate with material items for the child being in a single-family home.

I know many therapists use bribes to get kids to come to their office—candy, soda, or rewards. Other therapists allow the kids to sit outside at tables, or play sports during their therapy session. I myself have never relied on these machinations to get kids to my office. I always conduct sessions in my office (unless observing a child in class for some reason such as an upcoming IEP or the child is on LOP). Sitting outside only means that the child will be distracted by peers, more focused on the nearby sports activities than listening to our conversation, and other students will sit down and interrupt us.

I do not play sports with them, because they get to play sports during physical education, and counseling is not PE—it's that simple. It's hard to talk about frustration tolerance, bullying, or drug abuse, when the child is racing around shooting hoops, yelling, "Did you see that?"

When I first went to get Shanaya, she came bouncing out of her class all smiles. "Goody, I get to go to counseling. My counselor at my old school used to give me treats. Can we go to 7-11?" I stopped her right there, with the door to the class open so the teacher could hear what I was saying. I explained to Shanaya that coming to counseling was a privilege, and that based on that privilege she was getting out of class and not doing class work for an hour. I explained that I did not give students treats or rewards, but I expected them to behave in my office, or I would bring her back to class.

"Now, you can come with me and have fun playing games or doing arts and crafts while we talk, and you accept these rules, or you can sit in your class and I'll observe you and that will be your counseling time. Which is it?"

Shanaya stared at me with big eyes, and looked like she was going to argue, then capitulated, saying, "Okay, lets go." She then had to add on a mumbled sentence about me being mean and tough, but I had been called worse things than that. Nowadays when kids tell me I'm mean or tough, I just say, "That's right. And don't you forget it."

As I said, the children discuss their counselors. Several of mine have said to me, "You're not like other counselors," to which I always reply a neutral, "how so?" or "Is that a good thing or a bad thing?" Invariably, they tell me that I can't be manipulated. If the kids know which counselors can *not* be manipulated, that means they also know which counselors *can*.

The reality is, the perception of being tough or strict is not a bad thing. When I had been working with Jesus, a seriously depressed gang member for about two years, he made the exact same comment to me. He mentioned that I never made a contract with him, or took him out to buy stuff. I asked him if this bothered him. He said, "No, it's not that. The other kids in the class, they always come back from counseling with food and stuff. I don't get it. They're supposed to be *working* in counseling. Not going out and playing around. They don't have to pay to see you. If they were in the real world they'd have to pay to see a therapist. They're acting like little children, wasting their time. I don't get it."

10

Transference & Countertransference

Transference is defined as when a client reacts unconsciously towards you, either negatively or positively in a manner, as he would to a person he or she is emotionally close to. Countertransference, therefore, is when the clinician (or teacher, childcare worker, etc.) reacts toward the client based not on the client's objective behavior, but because they remind them of someone from their own personal life (Corey, 1996).

The problem with working in such unstructured environments as schools, residential facilities, or group homes, is that the more contact you have with your client, the more likely transference and countertransference issues are going to come up. This is why, from a clinician's point of view, it is so important to be aware of what your personal issues are. If you are not a clinician—but are a teacher, childcare worker, or aide—and find yourself feeling unusually hostile or attached to one particular client or student, you may need to step back and look at the picture objectively.

Ask yourself the following questions—who does this student remind me of? Is this student really any different from the other students? Does his or her behavior warrant my special attention? Different reaction? A good check is to ask yourself these questions—is my behavior in the best interest of the client? And is my reaction to their behavior in proportion to their behavior? If the answer is no to either of those questions, you may want to consult with a colleague at the very least. If that does not help, then it is time to consider seeking professional help.

I had experienced transference in private practice—but nothing prepared me for my first case in a setting where you see your clients five days a week. At the school in Watts, Guillermo, who was eleven, one day lashed out at me, asking me if I was a lesbian. I was completely surprised by this question, and asked him why he wanted to know. He then burst into this angry tirade about having seen me and a female colleague leave the school repeatedly together (we were picking up lunch for the therapists to eat during supervision). I explained to him that she and I were just friends, and I asked if it would matter if I was a lesbian. Guillermo seemed too young to be having "crush" feelings for me, and I was right.

It turned out that Guillermo had concocted an elaborate fantasy in his head where I would marry his one-to-one aide, and we would become his new parents. Once this was clarified, I was able to talk to Guillermo about the issues he was having at home, and explain to him that I would always be his therapist, but unfortunately, that was my only role. That didn't put an end to Guillermo's attempts to match-make between me and his aide, nor to the jealousy he displayed over my friendship with my co-worker. Clearly he had been building his fantasy world for a long time, and he had a hard time accepting the limitations of our relationship. He eventually lashed out at the other therapist, calling her names and telling her to keep away from me.

It was also at Watts that I experienced my first case of intense countertransference. I had a young girl, Lydia, on my caseload, who was creative, intellectually above her peers, and very sensitive. She loved to read, and would read or do art work almost every week in my office. After working a few months with her, I announced in supervision that if I were to ever adopt a child, I would adopt someone like Lydia. As soon as I heard the words out loud, I knew I was screwed. The problem with working in a school environment, is that when you work with the students in your office, they are with an adult one on one. They are generally calm, well-behaved, quiet, and focused. If you observe them in class, their behaviors range from off-task and distracted, to manipulative, bullying, verbally abusive, and sexually inappropriate. It is easy, as a therapist, to feel betrayed by your client when you witness these behaviors. You *know* they can behave well because you see them do it week after week in your office. Yet when you see them in class, their behavior deteriorates. Of course, in your office, there are no distractions, no peers instigating them, no difficult academics to frustrate them, and they are not being made to do activities which they would prefer not to do. In other words, in your office, they are having fun (generally). So it becomes easy to personalize it when you see a child misbehave and feel as though it is your failure when your client acts out. You feel as though your client has betrayed you.

Such feelings went through me when I was in the administrative office, and Lydia was there receiving her medication. She was throwing a tantrum, and using profanity, and I overheard her. I got in her face about it, telling her off for her behavior. Normally, I would respond to a child by calmly saying something like, "Hey! Apologize for speaking like that." But I let her have it! I was mad! Why? Because I thought the relationship we had was special. It was special—to me. Why? Because Lydia reminded me in many ways of myself as a child. Very smart, very emotional, a perfectionist, creative, and a little too bright for her own good. After that, I had to step back and detach from her. That does not mean that I stopped caring. It just means that I cared for her the same way I cared for everyone else on my caseload. No more, no less.

When you work around adolescents, that means one thing. Raging hormones. Eventually, there is going to be sexual transference. Feelings of projected love. When I worked at the group home, a young man, Frank, came in at age twelve. Technically, our group home took boys as young as twelve, but the entire time I worked there, most of the boys were 15–18. As a result, when Frank arrived, the older boys wanted nothing to do with him—he was the annoying, younger brother that they had nothing in common with. As a result, on the days when I was at the home (and I was there for hours), he would follow me around like a puppy dog. I took his behavior for what it was—initially. He was lonely and was having trouble fitting in with his peers.

Frank ended up leaving for a few months—I can't honestly remember why. I think he visited relatives. Frank came back a year later. When he returned, he had changed. He was more violent and psychotic. At one point, I had to call the PET team on him. He was threatening to harm a staff, the property, and was delusional. Of course, by the time the PET team arrived, he was calm and lucid. When I saw him the following week, he came into the office to meet with me, and said, "I'm so mad at you! You had the PET team called on me!" Then he twirled around several times, looked at me with a grin, and said, "Aw, I can't stay angry at you. You're too pretty."

Within a week after the PET team's refusal to remove him from the home, Frank had assaulted someone in the home and broken a window, and was to be transferred to another house. We had our final session together, and Frank asked, "Since it's our final session together, can I kiss you?" After explaining to him, gently, why that was not appropriate, I then went on to explain to him that eventually he would fall in love with someone closer to his own age, and that his feelings for me were perfectly natural. I told him that I was flattered, and that some day he would find someone who returned his feelings. Frank seemed frustrated by my explanation, but accepted it. I then documented everything verbatim in his notes, called the group home manager, and without violating confidentiality, explained that one of the boys had wanted to kiss me, so if any of them accused me of sexually molesting them, it was a lie. When you deal with this population, it's always good to cover your ass. The group home manager just laughed, and said, "I can guess which one it was."

Of course the ironic part of the story was that Frank did not get transferred to the other house on time, and had to see me the following week. He was extremely embarrassed, and did not want to discuss what had happened. Given that he was leaving the home, I didn't press it. He was barely able to give me eye contact. I just said, "Look, I remember what it's like to have a crush on a teacher or someone much older than myself—we've all been there. Don't worry about it. It's perfectly normal." I figure, these kids are always getting told how they are *not* normal, any chance to point out how their behavior *is* normal—I'm going to jump on it.

Group homes are a hotbed for counter-transference issues with the childcare workers. These employees are with the children, often for 8–12 hour shifts, and when a child is in the group home for year after year, it is easy to get attached to the child and personalize their behavior. I taught monthly in-services at the group home, and one month I discussed transference and counter-transference. One of the staff members brought up Ken, our resident drug-user. She asked, "Is this why I get so angry at him? Every week he swears he's going to stop using, and then not even 24 hours goes by before we catch him stoned out of his mind. I swear I want to shake him to death! What is that child's problem?"

I looked at her with raised eyebrows, "He's a drug addict?" Then I asked her, "Is there anyone he reminds you of?"

She thought about it, and clapped her hands to her face, "Oh, my, yes. He reminds me of my sister's oldest son!" She then went on a tirade about how this relative had caused problems for the family and what not.

I put my hand on her arm, stopping her. "Ken is not your nephew. He's here for a reason. If he had a supportive family, and hadn't been born into a drug-addicted family, he might be able to stop using. But right now he doesn't want to. And not you, nor I, nor the group home manager, or anybody else sitting around this table can make Ken want to stop using. Until he makes that decision, he's going to keep behaving the way he does. He is not doing it to annoy the staff. He's just doing it because it's the only coping skill he's ever known. Don't take it personally, or every time he walks through that door loaded it's going to cause you stress and you'll stop wanting to come to work. Believe me, it isn't causing *him* stress."

The rest of the staff members then had a lively discussion about how it did feel like Ken's continual breaking of his promise to "not use" felt personal. When you work with someone so closely, and witness them slowly destroying their life, you can begin to feel helpless. That helplessness can turn into anger, and anyone working with these kids needs to be careful not to turn that anger onto the child.

At my current school in the Valley, I see a lot of my students have transference issues not only with me, but also with their staff. I had a high-functioning girl, Kelly, who was bi-polar. Academically, Kelly was on track, but she was in school due to her behavioral problems and a lack of medication compliancy. She did very well during summer school, and had no problems, but when the fall term started and she had a different teacher, Kelly started acting defiant. She would often call her teacher a "bitch," "stupid whore," and "fucking piece of shit." Kelly would sit in my office, and say, "That's stupid bitch can't tell me what to do. I don't have to listen to her. All she does is talk in my face. It's annoying."

After listening to this for a few months, I asked Kelly one day after another tirade against her teacher, "Does Miss Jessica remind you of anyone?" Kelly sat there for awhile thinking about it. Then she sat forward, slamming her hands on my desk. "Yes!" she shouted, "She's just like my mother!"

We spent the rest of the session talking about ways that Kelly could leave her issues with her mother at home when she came to school, and learn to see her teacher for who Jessica was. Yes, the teacher was going to tell Kelly what to do—that was her job, but Kelly needed to stop taking all her anger at her mother out on her teacher. Eventually, Kelly grew to respect Jessica as a teacher, and while they still had their problems—they were primarily due to Kelly's bi-polar disorder and the associated mood swings—not because she was attacking Jessica as the most nearby object representing Kelly's mother.

Another young black man that I have worked with for over four years—he was twelve when I started with him, lives with his aunt and uncle. He has never known his biological parents. Jake considers his aunt and uncle to be his real parents, but that does not mean he does not have a lot of issues with them. They let Jake know repeatedly that it is a "burden" raising him, and that they were done raising their own kids and didn't need to take on another child. I have watched Jake grow from an irresponsible, angry kid who would have to be physically restrained as he regularly got into fights, to a mature young man who the staff trusts with additional responsibilities.

However, Jake continues to be immature at times, and I joke with him, "You know, I have a two-by-four with your name on it." That's what I tell my kids when they become obstinate and stubborn. I'll tell him, "You want to be treated like an adult, but you think you can run around and play without any consequences. Let me tell you what being an adult is—it's about accepting responsibility for your actions. If I don't come to work, I don't get a paycheck. If I don't get paid, I can't pay my rent and I get evicted. You think you can sit in class and goof off, but then you complain you aren't learning. Whose fault is it if you aren't learning? I see your teacher doing her job, but you aren't doing yours." Jake and I have a talk about taking responsibility for his actions probably once a month. I lecture him until I feel like I can make a tape-recording of myself and simply play it. I do not actually need to be in the office, and I've told him that. But every time he sees me, he shouts out, "Hi Miss Jillian," and he gets upset if I don't acknowledge his hello.

Jake loves to play dominoes. We've been playing that since I met him almost five years ago. When we first met he could never win. Then he gradually started winning, and now sometimes he wins all the games. One Monday I picked him up after Mother's Day, and he was losing every game. I asked him, "Are you throwing the games? Cuz something ain't right here. I'm winning way to easy. If you aren't going to play to win, I don't want to play. I don't want a cheap victory."

Jake grinned, and said, "Aw, but it's Mother's Day, Miss Jillian."

"Mother's Day was yesterday, and I'm not your mother."

"Yeah, but I didn't see you yesterday, and you're just like my mother."

I have many kids on my caseload who will tell me that. Especially after I lecture them and get right in their face. These kids want to know they are loved with structure—not with bribes or candy. They want to know you care that they do their best. When a kid misbehaves and I say I'm disappointed in them, they'll say I sound like their mother—not in a negative way, but in a positive way. For me, that is the best transference, and best compliment, one of my kids can pay me.

Perhaps the most complicated case of transference I had to deal with was Myles. Myles came from a divorced family. However, his parents maintained their dysfunctional relationship after their divorce, often putting Myles in the middle (a psychological term referred to as triangulation). Whenever Myles' father was upset with his ex-wife, he would attempt to get back at her by becoming the "preferred parent" and promise Myles something, such as a trip to a Dodgers game. Of course, he and his ex-wife would make up, and the promised trip would never transpire. Myles' mother would do the same thing, and when she was angry with her ex, she would make some big promise to buy Myles some expensive present that she would then fail to keep.

Poor Myles spent three years coming into my office, excited, telling me what great event was going to occur in his life, only to come crashing down bitterly the following week in tears when the promised activity never took place. When Myles was old enough to understand how he was being used, we worked on his learning to extricate himself from his parents' manipulative behaviors, and not to fall victim to their empty promises.

However, in school, Myles was unused to having two parental figures who were united and in agreement. So Myles would continually try to sabotage the relationship I had with his teachers, and vice versa, because he did not know how to handle two adults who actually were on the same page. He had never been "parented" by a unified front, and he was more comfortable by chaos. Myles would frequently tell me lies about what his teachers had said about me, but I would never buy into it. I kept telling him that in his session, we were there to talk about him—not me, and not his teachers unless it was a problem that *he* had with his teacher. He would say, "But doesn't it bother you that Sam said that about you?"

"Nope. But if it bothers *you*, then we should go to Sam right now, and the three of us should discuss this together."

Myles would look horrified, and say, "No, no. I don't want to talk to him about it. I just thought *you* should know what he said about you. If it was me and someone was talking bad about me, I'd want to know about it."

"Really. Well, it sounds like this is really upsetting you. Maybe we should go talk to him about it, then. Because I don't want you to be upset."

At this point, Myles would be more and more agitated. "No, no. It's okay. Really, I'm not upset. Let's just play poker, can we? Just drop it. Okay. Forget I brought it up. I don't want to talk about it anymore."

What Myles did not know is that I talk to each teacher probably one hour a month about all the students on my caseload in their class. I have a pretty good idea of how the teachers feel about me, because they don't try to hide their feelings. Every teacher of Myles had come to me begging for advice on how to deal with him, because of his many problems, and the difficulties with his family. Students often end up in Special Education because they are used to chaos, and in a normal environment where no chaos exists, they will create their own chaos, because it is what is familiar. Myles was unable to handle parental figures who didn't argue over him, and therefore he felt the need to create an atmosphere conducive to argument. As he matured, this behavior eventually stopped. Did I confront Myles directly on his behavior? No. That level of discussion was too mature for him to comprehend, and I felt calling him out on his lies would be too shaming. By giving Myles the opportunity to have a three-way discussion, and him turning it down, I made my point in a more subtle, but less shameful, manner. If I had listened to his "gossip," I would have reinforced the behavior, and it would probably have continued.

The young man, Andy, I mentioned in earlier chapters, who was to leave for public school after working with me for over four years, struggled to deal with his feelings for me the last month before he left the school. Since I had been the one therapist not to abandon him, Andy did not know how to say good-bye in a healthy way. He spent his last month in session finding ways to get angry with me so that the ending of our relationship would be less painful. Sometimes, transference can have both a positive and a negative cost. It can help the child to grow, but it can also hurt when they have to leave the nest.

11

Oops! Have I Said Too Much?

Just as countertransference issues can intensify by working in an environment with no physical boundaries, so can the tendency on the part of the professional to disclose too much personal information. In some ways, self-disclosure when working with youth can be an excellent way of establishing a therapeutic relationship. It can also be a great source of education for kids who tend to live in a world that is rather myopic. However, it is very easy to cross the line from self-disclosure, which serves a therapeutic purpose, into a territory where damage is being done. How then, is a clinician to understand the difference?

As mentioned earlier in the chapter on setting boundaries—ask yourself these two simple questions. Does this serve a therapeutic purpose? And is this in the best interest of my client? When self-disclosing, the answer should be yes to both of these questions.

Because I work at a school where many therapists see their kids on the playground, sitting in the library, or school cafeteria—or for whatever reason they do not even bother to close the door to their office while in session—I have overheard a lot of conversations in my years. Many of the self-disclosures I have heard from other therapists have been inappropriate. One particular therapist whose office is next to mine, had a child who was very loud. The louder the child got the louder the therapist got, and it was therefore easy to hear entire conversations, despite my repeated knocks on her door and requests to be more quiet. So here are some following examples of when self-disclosures are clearly not appropriate.

A therapist, we'll call her Joan, was with a child who had noticed the picture of her Joan's son on her desk. The child, we'll call him Billy, asked Joan about her son. Joan went on in great detail talking about her son, what grade he was in, what he enjoyed doing, the kinds of sports he enjoyed playing. (Is talking about your family therapeutic? Is it in the best interest of the child? Maybe, if the child's IEP goal is to work on positive social skills). Then Billy asked if Joan was planning on having more children. Next thing I heard was Joan talking about how she could not have any more children, and suddenly the therapeutic roles were reversed. Billy was comforting the therapist telling her, "That must make you sad. Your son must be lonely." Again, unless the client's IEP goal in counseling was to work on his positive social skills—and there must have been other ways of doing so without revealing so much personal information, this was completely inappropriate.

By revealing too much personal information to kids with emotional disturbances, you can often recreate their family environment where these children are often the parentified child and they feel responsible for taking care of their siblings because the parents are too intoxicated or overworked to do so. In this scenario, Joan may have recreated the family environment in her therapy session by allowing her client to become the parental figure comforting the therapist.

Additionally, self-disclosures need to be handled carefully when working with emotionally disturbed children. Do you as the clinician want all the kids in your client's class to learn you can't have children? How about on your client's bus? What if Billy uses this information to hurt you, the clinician, because he knows something about you that gives him power over you on an emotional level?

I have a simple policy with the kids I work with. I tell them that I will always tell them the truth when they ask me a question. I tell them this because I know that they are used to adults lying to them. Parents, social workers, probation officers, teachers, staff, previous therapists—not all—but most, have lied to them and they assume the same will be true of me. It takes the clients assigned to me usually one year before they believe that I will never lie to them. However, the caveat to this is that if they ask me a question that is inappropriate, I will tell them so and not answer them. *But I will not lie to them.* Questions that are inappropriate are related to my salary, my personal life, my politics, and my religion. The reason for this is these topics are not therapeutic or educational. However, because many of my clients are low-functioning or ignorant of how the real world works, sometimes self-disclosure can be both therapeutic in that it confronts their behavioral issues and educational in that the client learns something they probably would not be exposed to otherwise.

When working at the group home, one of my clients who was to enter college soon wanted to be an entrepreneur when he graduated. He was going to have his own music line, clothing line, and who knows what all else. I worked in the entertainment industry for fifteen years, and have done the corporate bit. So I explained to him the difference between being self-employed, and working for someone else. I asked him, "If you go on vacation, who's going to pay for your vacation? If you get sick, who's going to pay for your sick time?" At the time I was working two jobs, just like I currently am. I explained to him about self-employment taxes, workers' comp, and disability. Again, I felt my client was allowed to make any choice he wanted, but these kids go to special schools, and do not have parents to guide them. If he wanted to make a choice to be self-employed he needed to have all the information before hand, and his social worker or the group home staff were probably not going to provide it to him.

A client at my school, Mike, kept complaining that the work he had at our school wasn't hard enough, that he wanted to return to public school and get a "real" education. I pointed out to Mike that most of the times I saw him in class he was not seriously doing his work, but was just doodling on his papers. I asked him if when he was assigned to write in his journal he wrote just what he was asked to write, or if he wrote more than the minimal requirements. I then went on to describe to him what I had been required to do when I was in school. I discussed in vivid details how long my English papers had to be and that they had to be typed. I described my science project that involved collecting 50 different types of bugs and labeling each one. I told him about having to read multiple books—not just one—and then give oral reports on the entire subject, not just one book. Mike stared at me in horror. I said, "So until I see you do the work you are assigned here perfectly, and take it to the next level on your own, don't even think about returning to public school."

A manic client of mine who was ready to move out on her own was anxious to get her first credit card. She kept talking about how she had "needs." I explained to Sharon that she tended to confuse her wants with her needs. It turned out that one of her needs that she was going to buy with a credit card was a car! So we talked about insurance, gas, smog inspections, and vehicle registration costs. We talked about how credit cards work, and interest rates. Sharon was hoping to get a job in retail, and I suggested that if she did that, she put a specific amount of money away each week in a savings account, and build that up for an emergency fund, instead of relying on an emergency credit card. I then gave her an example of how my credit union works, and how I can have money put into separate types of savings accounts that I can't have access to until a certain time. I gave her examples of how by doing this I was able to buy a car one time with half the down payment in cash, and another time take a vacation, again all in cash. I explained that it took a little more patience, but cost less in the end because I didn't have to pay off any interest. Her response after all this financial conversation was, "Damn, miss, this is too much responsibility!"

Were these self-disclosures therapeutic? Yes, I think so. My clients learned from the information. Was it in their best interest? Again, I think so. When the information has to do with issues that the clients bring up related to their *behavior*—skipping college to pursue a pipe dream career, failing to complete class assignments because they aren't "real" class work, and manic spending and lack of financial responsibility for one's impulsivity—self-disclosure can be an excellent tool to connect with the client and motivate them to change. Self-disclosures can be especially powerful if the therapeutic relationship has already been established and the client sees you as a role model.

Another example of a form of self-disclosure—not even self-disclosure, but a session that violates both the laws of state and the Board of Behavioral Sciences (the governing board regulating marriage and family therapists), is the following. A therapist, we'll call him Tony, was in session with a girl who was particularly loud, we'll call her Shatoya. Shatoya came in and asked some question about God—I really do not know what it was, because I was at my computer doing paperwork. Tony's correct response would have been to either explore Shatoya's *feelings* about God (feelings are always therapeutic), or if Shatoya had specific questions about religion, refer Shatoya to her church's minister or priest. The reason for this is, there is a law separating church and state, and since our school is not a private religious school, teaching religion is against the law. Secondly, there is a law among therapists called "scope of practice"—unless Tony was a minister, he was practicing outside of his scope of practice. Either way, I seriously doubt there was anything in Shatoya's IEP counseling goal related to religion. Normally, things worked on in counseling relate to anger management, frustration tolerance, bullying, or family problems.

With all that in mind, I continued to hear Tony spend the next half-hour teaching Shatoya about the role of God, Jesus, Mary, Joseph, and the bible. Also about the role of heaven and hell. Now, here comes the tricky part. Let us say Shatoya goes home that night and says, "Mommy, did you know that Joseph is Jesus' step-daddy?" Best case scenario, Shatoya's mother asks her where she learned this, finds out her counselor at school told her, calls up the school, and demands another counselor for her daughter. Worst case scenario, the mother is smart enough to file a complaint with either the BBS and/or the school district which could result in legal ramifications.

For therapists with no boundaries, it is very easy to cross the line, and consider the children at the school your friends. There is nothing wrong, as I said, when I child asks me what religion I am, in answering, "We're not here to discuss me, we're here to discuss you." If that answer offends them, and they whine, "Why won't you tell me?" I tell them I have a no religion/no politics rule—that it's not the best use of their therapy time (I used that line *a lot* during the 2008 election). When that still bothers them, I just say firmly, but with a smile, "When I'm in your house, I will abide by your rules. But you are in my office, and you have to follow my rules."

Things I *will* disclose to my kids—that I work multiple jobs. I feel it helps them to understand why I am not at the school five days a week and why my hours are erratic. Also, it is helpful for them to appreciate that adults often have to work two jobs to support themselves. It will better prepare them for their life. If they ask me what my other job is, I let them know. Since my other job involves writing reports, I hope this encourages them to better their reading and writing skills if they understand that many jobs require these skills. I will disclose that I wasn't perfect as a teen, and that I went to three different high schools in three different states, so I understand what it is like for them to constantly change schools. Specific details about how I wasn't perfect I leave out. I just let them know I got in trouble, and that my parents and I didn't get along either when I was their age, but now we do. Vague information is enough for bonding—specific information makes the session become about me, not about them. The reality is, once I tell them the vague history, they rarely ask for more. I've given them enough to satisfy that I'm cool and being open with them. Deep down, if you give these kids too much information, they lose respect for you. As was the case of John in the previous chapter, when a client does spend his entire session asking too many personal questions about you, that is a sign of transference and needs to be investigated and addressed in an appropriate manner.

Many of my older adolescents will complain that their parents tell them too much personal information. Especially the kids in single-parent homes. The parentified children. They complain that their parent come home from work and talk about conflicts with co-workers, problems with relatives, their financial stresses. These kids already get "too much information" from their parent. As one client said to me, "I don't need a friend, I need a mom." With this type of dysfunctional family life, the last thing they need is too much information from their therapist.

Bonding is one thing. Turning the child into your therapist, friend, or confidant is just plain wrong.

12

Show No Fear

These are children who have often been in the system since birth. They may have been removed from their parent due to the mother being on drugs, being a prostitute, or they were removed from the home as an infant due to neglect. The mother may have been declared by the state to be mentally incompetent. Often these children do not know who their biological father is, or he is in jail. As a result of these children being in the system for such a long time, by the time they reach adolescence, they often know how the system works better than their teachers and therapists.

Therefore, they can detect an inexperienced professional and take advantage of them. As mentioned in the chapter on boundaries, these are kids who can often manipulate inexperienced professionals in order to obtain what they want. Then on the other end of the spectrum are the ones who work on a different level—fear and intimidation.

I can always tell when a new teacher is at our school, because the noise level in a particular classroom will suddenly escalate. The kids will be screaming, out of their seats, throwing things, desks will be overturned, and students will be running out of the class with a staff member in hot pursuit (better known as AWOLing). But that is not the only noise. There will be the continual sound of yelling—by the teacher.

Then there are the classrooms that are taught by the teachers who are experienced and in control of their classes. They are quiet, the students are on task, they raise their hands for permission to speak, the students are facing the direction they are supposed to face (forward), and when they are listening to the teacher they are attentive. When there is an out of control student, the student is quickly contained, and everything returns to normal.

Teachers often appear to equate volume of speech with control over their classroom. Therefore, they yell and scream to get their students in control, and slam their hands against their desks to get the class' attention. Unfortunately, what these teachers do not realize is that they are recreating the students' family environment, and thereby losing their students' respect.

When I speak to my clients, one of the things they often complain about is how they dislike their family environment because of all the yelling. When we talk about their relationships with their parent(s), what they tell me is that they dislike the fact that their parent never listens to them. All they do is yell at them. The clients feel that their parent's do not value their thoughts, feelings, and opinions—the students just become a repository for their parent's anger. Or if the child and parent has a conflict, the parent is not interested in hearing the child's version of the story, but dismisses them by yelling and screaming, and to put it in their lingo, "they get in my face."

When working with hard-to-reach adolescents, it is imperative to show no fear. Once you have let your clients or students know you are afraid of them, you have lost control and given it to them. In a classroom, yelling shows that you, as the teacher, are out of control. I have seen teachers control their classroom with a look, a count to five, or a single word. The size of the teacher does not matter. Two of the best teachers at our school weighed less than 100 pounds, and both were women. One of them taught elementary grades, one taught high school. However, the teacher who taught elementary grades was able to handle the big kids just as easily as her own.

One day in the area where the students gather every morning after the buses unload, my borderline student, Shanaya was out of control, and physically acting out. Her teacher had asked her repeatedly to sit down, but Shanaya was doing what she wanted to do. I stood nearby, watching, because I wanted to be able to discuss with my client her behavior in her session. An aide came out and redirected her, and Shanaya started screaming, "I don't have to do what you tell me, you fucking bitch!" and she became physically aggressive.

Next the aide tried to physically redirect her, and again Shanaya was physically aggressive and swearing to the hilt, saying things like, "Get your fucking hands off me you fucking lesbian!" The aide who was redirecting her probably was several inches taller than Shanaya and weighed a little more than her. The aide gave up and walked away, and Shanaya was swaggering, swearing after her. At that moment, diminutive, Jennifer from the elementary school came up and grabbed hold of Shanaya from behind, saying, "That is enough!" and strong-armed her to her class, giving her a time out. My client probably weighed forty pounds more than the teacher.

When I went to follow up on my client, Shanaya was sitting in her classroom with one of the classroom aides. She was still swearing and carrying on, but she was sitting at a desk, and she was crying. From that point on, I would joke with her that if her behavior ever got out of control, I'd just send her to the elementary school, and she could sit in Jennifer's class. Her response would be, "That's okay, she's cool."

I cannot reiterate enough that these students live in home environments where adults are out of control. The parents have drinking and drug problems, employment problems, domestic violence issues, or are sexually promiscuous and strange men come and go making it hard for these kids to trust adults. They have no adult role models that they can respect. If their teacher, who should be a positive role model they can respect, appears afraid of them and act out of control by yelling, screaming, shouting—all the world history, algebra, science, and literature knowledge in the world will not earn their respect. These children are not awed by facts and figures, they are awed by your ability to help them with their behavior. If these children are acting out of control, it is because their environment at home is out of control—recreating that environment will not help them. It is only after you have earned their respect that you can educate them about other matters.

One week I was in such an out-of-control class observing two of my students while they were doing the mandatory state testing. As a result, the therapists were not able to take their clients to counseling because testing was going on during the majority of the school week. One of my clients who had tendencies to bully his peers, refused to do the testing. He went through the test answer sheet without opening the book and circled in answers randomly and then asked the teacher (who was new) for some paper to do art work with. She gave him the requested supplies. The other students in the class saw what had happened, and they all proceeded to quickly fill in their state testing, and request art supplies.

I spoke to the teacher outside and pointed out to her that by rewarding my client's off-task behavior with a preferred activity, she was encouraging him to continue to behave in a similar manner in the future. She whined, "But if I don't give him what he wants, he'll disrupt the class." What she failed to see, was that he disrupted the class by causing the others to also be off-task when they saw that he was being rewarded for his negative behavior. If she had given him a negative assignment to do for failing to complete the testing, the other students would have stayed on task in order to avoid a similar punishment. The reality was, the teacher was afraid of her students, and they knew it. She was afraid to punish them—she just wanted them to be quiet and engaged in some activity rather than "creating a scene."

I had a high school student who had just gotten in trouble that day, and was angry with his teacher. I picked Martin up and all the way to my office, Martin was ranting and raving about how he wanted to kill his teacher. If he ever got his hands around the teacher's throat, he'd choke him to death. I asked him, "Are you just venting, are is this a serious threat?" The student did not understand the word venting, so I had to clarify my question for him—but I did praise him for admitting he did not understand a word I used.

Once in my office, we discussed what had happened in class to make Martin so upset, and how he could handle the situation differently in the future. However, Martin was too angry to discuss coping strategies or conflict resolution skills. He just wanted to be out of class and away from his teacher. Every time I brought up how there were rules in school that he needed to follow, just like there are rules in society that he will need to follow as an adult, Martin would go off, talking about how he didn't care, and the rules were stupid, and the rules didn't apply to him, and he didn't have to follow the rules if he didn't want to.

It was apparent that Martin did not want to work on his issues, and was using his therapy time to get out of class, so I stood up, and said, "Come on—we're going back to class. This is a waste of my time."

Martin looked at me. "Are you serious? What'd I do? Are you afraid of me?"

I gave him an incredulous look. "Do I *look* like I'm afraid of you?"

He shook his head, "No."

"Exactly. I'm *annoyed* with you. There's a difference. You broke a school rule. I'm not going to reward you by getting you out of class. You have a problem with your teacher. Go back to class and work it out with him. We're through here." My client was angry with me, but followed me back to class. I'm only five feet two inches and weigh less than 120 pounds, but my students know better than to mess with me. They also know I'm not afraid of them.

New students at the school will always test you, especially if they are older and have been shunted around from one placement to another. I was assigned a student who had been in juvenile hall, and his previous therapist felt she had no connection with him and a new therapist might be able to help him better. When I went to observe him the first week, he was in class causing total chaos. All the other students were redirecting Miguel, telling him to sit down, follow staff's directions, calm down, stop using profanity. It is a bad sign when the students start redirecting another student. However, this was a high-functioning high school class, and Miguel's behavior was disruptive.

Apparently Miguel felt he was special. He was demanding to speak to the principal, the IEP coordinator, and various other administrative staff. Why? Because he did not want to do his assigned work. Eventually he demanded to see his counselor. I raised my finger, and said, "I'm right here." I explained to him that his previous counselor would not be working with him anymore, and that I was his new counselor.

Miguel asked to see me. I asked him, "Are you asking to see me so that you can discuss whatever is bothering you? Or are you asking to see me simply to get out of class and avoid doing your class work?"

Miguel danced around the question, not answering it, and again the students redirected him. Then he started using profanity. I kept waiting patiently for an answer, and eventually he said he wanted to come with me to talk. I knew that he was lying, but I was willing to give him enough rope to hang himself.

As soon as Miguel and I were outside the door, he said, "I should AWOL."

I shrugged, and said, "No one's going to stop you."

He looked at me, "What kind of therapist are you?"

"The kind who gets paid whether you AWOL or not. Let me ask you something. How old are you?"

Miguel refused to answer the question.

I continued, "I'm assuming you are old enough to know that if you AWOL, you have to deal with the consequences of your behavior. Given that you are on probation, the only one you are going to hurt by AWOLing is yourself. Not me."

Outside my office, Miguel started muttering at me, and calling me a "white bitch." I stopped him, saying, "Do you want to come into my office? Because if you do, you need to apologize for your language. I have just met you, and you know nothing about me. You asked to see me, and I'm seeing you, even though you clearly have an attitude problem. I have not treated you with disrespect. I have not called you racial names. So don't take your anger out on me." Surprisingly, Miguel apologized for his behavior.

Once in my office, Miguel started making demands to use my phone, just as he had in his classroom. I told him he was not in a hotel, but a school. We were not here to cater to his every need, and that he needed to get his act together and act his age. I told him I did not appreciate his coming to my office on false pretenses, and that I expected him to discuss his problems, not make demands. Miguel AWOLed from my office, and I did not run after him, although I reported his AWOL to the administration and his classroom. Unfortunately, Miguel was to disappear to the Halls a couple weeks later—no doubt in violation of his probation, so I suspect that getting even an apology out of him was probably a miracle.

Possibly the dumbest thing I ever did (in retrospect), in terms of showing no fear, was the following. I had been working at the group home for well over a year, and one of my clients who was very needy, and would have spent three hours in session with me if allowed (he was an example of indiscriminate attachment), was lying on the couch of the den watching television. I had just come out of the bathroom, and he looked up at me, whispering, "Jillian, you want to know a secret?"

"Sure, Jose, what is it?"

He motioned for me to come closer, "You have to promise not to tell. Okay?"

"You know I can't promise anything. Not if it's a danger to yourself or others. What is it?"

Jose started getting agitated, "No, no. You have to promise. I need it for protection."

Now I was getting concerned. "What do you need for protection?"

Jose lifted up his sweatshirt to reveal the butt of a gun tucked into his pants. Without even thinking, I held my hand out and demanded, "Give it to me!"

Jose looked at me with pleading eyes, "I need it for protection, Jillian. If someone bullies me." After several minutes of arguing (or maybe it was seconds—time becomes surreal in moments like that), Jose handed the gun over. Then he giggled manically, adding, "It's not even real. What's the big deal?"

I took the gun away, which looked real enough to me, and gave it to the group home staff. Turns out, they had confiscated the "toy" from him several times, but he kept getting it back. I suggested that they remove it from the premises. Then I had a long talk with Jose about how walking around that particular neighborhood with a fake gun was a guarantee to get him shot with a real gun.

Driving home from work that night in the dark, the adrenaline rush kicked in, and I thought to myself, "what the fuck was I thinking?" There was staff at the group home that I could have called to handle the situation. But in that moment, all I could think was—my client has a gun and I don't want him to get hurt. I have worked at an agency where a patient went berserk, threatened a clinician, and we had to lock the facility down and call the police. I was instrumental in handling the situation—I called the clinical director to let her know what was happening and I made the 9-1-1 telephone call. With my job working with a psychiatrist I have to evaluate patients who often are psychotic or suicidal and need to be assessed to determine if hospitalization is required.

However, in this one instance my instincts took over and I quite literally acted without thinking. Was it in the best interest of my client? Yes. Could I have handled it another way by calling in a group home staff member? Yes. Would my client still have respected me? Probably. On the other hand, I'm fairly certain, since there is nothing else to do at a group home than sit around and gossip, that all the other boys found out about what I did, and it just confirmed what they already knew. They couldn't mess with me.

I should point out here a note on safety. Just because I let my clients know that I am not afraid of them, does not mean that I endanger myself by seeing a student in my office who is clearly out of control. If I go to a class and a client is acting out physically, or their behavior is completely inappropriate, I will not take them to my office. I may go back to them later to see if they have calmed down and I will assess them at a later time. For example, one day I went to pick up my low-functioning psychotic client, and when I saw his face, I asked his teacher what happened. I was told that Deonte had just gotten LOP (loss of privilege) for being sexually inappropriate. Since Deonte was an AWOL risk, I asked him, "Do you want to come now, or do you want me to come back in an hour." He told me to come back. When I came back, the expression on his face was calmer, and he appeared lucid. I took him out of the class with his teacher, and asked Deonte, "If I take you to my office, will you behave?" He said yes, and I had no reason not to doubt him since he had always behaved with me and never lied to me. But the main reason I took him was his body language was more appropriate than when I saw him an hour before.

Another time, I went to pick up a student who had anger issues, and he made it clear that he did not want to come with me that day. This was a student who only wanted to come to counseling when it got him out of a class period he did not like. The student weighed almost 400 pounds. While walking to my office he punched his fist into a wall, breaking the wall (he had recently destroyed other school property), and so I turned around and said, "Come on, you're going back to class. You just earned yourself LOP." He then decided he did not want to go back to class and continued to head to my office by himself, and I reported him AWOL. Unfortunately, a lot of this work is instinctual, and you learn over time to read a student's body language. There is a lot of posturing and empty threats, and then there is behavior that is truly out of control. Over time you will either learn to recognize the difference, or you will quit the profession or go on to work with a different population. On some level it is like working with abused animals—you have to learn which ones will bite you, and which ones will let you pet them if you approach them the right way.

13

Spring Fever!

I fondly remember the anticipation of summer break as a child. Knowing that I would be taking a family vacation with my parents— usually to the Smokey Mountains where we went nearly every summer since I was around nine. Summers meant more time taking the boat out on weekends, going to islands where I could look for shells and explore the vast ruggedness of the wild terrain. And after Disney World opened, summers also included our yearly trek north to Orlando. Summers meant being able to stay outside later playing in the trees and yard. Or visiting my friends on my bike, and sleepovers when my friends and I would stay up late at night giggling under the covers, despite repeated admonishments to "go to sleep! I mean it!"

When I started working in the school system in Watts, I quickly learned that the school year with emotionally disturbed children has a different rhythm than it does for the rest of the world. It is cyclical and almost palpable in its predictability. It is like the changing of seasons in the north, when you can feel the weather about to turn. In Los Angeles, public schools are on what are called Track systems (unlike when I was a kid when all schools were on break at the same time). There are usually four tracks, A-D, and each track will be off from school at different times during the year. One of those tracks will mimic what is called the Traditional Year. So when working with students in an SED class in public school, their winter break could be in October, and their summer break could be in March.

The Traditional School year, which is what both the SED schools I have worked at follows, starts in September after Labor Day weekend. There is usually a three-week break around the Christmas holidays (winter break), and then spring break is around Easter for one week. There are two summer breaks—a short break of approximately two weeks prior to starting summer term, which begins after the July Fourth holiday, and then a longer break in August lasting one month.

For children who live in foster homes, or dysfunctional homes with parents who continually argue, abuse their children emotionally, or the kids have conflicts with their siblings which may get physical, these breaks—instead of something to look forward to—become something to dread. As a result, for two weeks in the school prior to an upcoming break, the kids start to act out increasingly more in anticipation of their break. As much as the students complain that they hate their school, hate their f-ing teachers, and hate the staff—the school provides a semblance of structure and normalcy that they do not get at home. Therefore, being at school is actually preferable than being at home for an extended period of time.

Jared, the young man I worked with for eighteen months before he returned to public school, was such a student. Due to the continued budget cuts in school funding, the Los Angeles Unified School District is always looking for ways to cut back on education. Summer school used to be six weeks. Then it was cut down to four weeks for elementary and middle school kids. It has once again been cut—but that's a whole other topic. However, one day in August Jared was unaware that as a middle-school student, it was the last day of summer school for him.

I went to pick Jared up for his session, and in my office, his body language was different. He sat slumped in his chair in defiance, staring into space. He wouldn't engage with me. When I asked what was wrong, he started shouting obscenities. "I hate this fucking school! It's fucking stupid! I don't care about fucking counseling! My teacher's a fucking bitch! You're a fucking bitch!" I just sat there, stunned, wondering what had happened to trigger this tirade. However, Jared had shut down emotionally, and nothing was going to change his behavior, so I took him back to class. After talking with his teacher, I learned that she had informed the class that it was the last day of school and they would not return until September. Suddenly a light shown through the clouds, and everything made sense. I wish I had been in possession of this information prior to going into session with Jared. I shared with the teacher how he had behaved with me, and explained that he was obviously nervous about spending a month at home with his family. I asked if he had ever mentioned any abuse issues. She reported that he had problems with his step-father, but nothing abusive.

When Jared returned to school in September, he was back to his old self. When Christmas break approached, I made certain to warn him of the approaching break so we could talk about coping skills over the holiday. Later, when Jared left for public school, I asked him, "Do you remember that one day in my office when you went off swearing and calling me a fucking bitch?"

He looked embarrassed. "Yes."

"Why do you think you did that?"

Jared didn't even need to think about it. "Cuz I was angry, miss. They told me it was the last day of school and I didn't know about it. I thought I had two more weeks before the break!"

Even worse than the summer, is Christmas break. If you are in foster care or a group home, Christmas is not a time of family traditions, presents, Christmas trees, decorations, and visits with family followed by a home-cooked meal. When I worked at the group home, there was a sad tree and a "county" dinner. Some of the kids are able to go on home passes, but this just makes it more depressing for those who do not have a home to visit. For children living with parents or extended family members, where the family environment can be difficult—Christmas is often a time of stress. There isn't money for presents, parents put things they can't afford on credit cards, alcoholics drink to excess, tempers flare, domestic violence increases, and gang activity escalates.

The most poignant story to demonstrate how Christmas is different for these children is the following. One week before winter break I saw Johnny, a little ten-year old with terrible self-esteem issues. He lived with his grandparents, and had been hospitalized many times for his "outbursts." I personally had never witnessed any of these outbursts, although I had conferred with the staff about them on many occasions.

The week before the break, Johnny was subdued in my office, begging to go back to class, which was unusual. I told him he could go back, after he told me what was wrong. Suddenly, he burst into tears, and said, "I'm afraid they're going to send me to jail!" After much questioning, I learned that his grandparents had apparently told him that if he misbehaved over the break he would be "sent to jail."

Whether or not this was true was irrelevant. What was relevant was that Johnny believed it. He confessed to having broken many items in his grandparents home when he was angry, so we spent the session discussing anger management tools and coping skills to get him through the three week winter break. I reassured him that at age ten, the worst thing that could happen to him was that he would be removed from the home, probably to a group home, but that he could not go to jail, and it was doubtful that he would be sent to Juvenile Hall over breaking things in the house. Johnny had a good cry in my office, then returned to his class. At no time in the session did he mention Christmas or anything that he was looking forward to. For him, the break was one long probationary period where he had to be perfect or else something bad would happen to him.

Another example of a different sort is Maria, my student who continually had gang-related deaths occur over the holidays. Prior to the winter break, we would always talk about how she would cope over the holidays if she started to feel depressed or suicidal. Since Maria had no outside support services, I encouraged her to talk to the minister at her church if she felt seriously depressed. I also gave her the teen hotline number to call. Students tend to lose information you give them in writing, so I prefer to provide a resource that is physical such as a priest or close relative—even if they aren't religious, because the priest hopefully will be able to refer the student in the right direction. We talked about a "safe" house that she could go to if she felt like she was in danger of getting into a fight with her boyfriend or her mother.

I often spend the entire week prior to breaks discussing such coping tactics with my students—particularly the older ones who are at risk for running away from home or turning to drugs or violence without the structure of school. We come up with plans of who they can talk to that they trust; where they can go if they need a "time out" from their family; and in a real emergency, where they can stay if necessary for the entire break that their family will agree to. Often their family is more than willing to cooperate and let an older adolescent go live for two weeks with a cousin, grandparents, or older sibling. This is especially important in single-parent homes with mothers and daughters, who tend to get on each other's nerves more often than fathers and sons.

After the winter break, when my kids return to school, I never ask, "How was your Christmas?" That question is too loaded. It often makes the child feel obligated to answer "good," in order to make *you* feel okay. I always ask, "How was your break?" with no reference to the holiday. This way, if they want to talk about Christmas they can, and if they don't they aren't under any pressure. I had a child who spent Thanksgiving living in a homeless shelter with his mother and siblings. How shaming would it be for this child to have multiple people coming up to him saying, "Hey! You have a great Thanksgiving?" That question almost demands that the client answer in the affirmative. He was trying to hide the fact that he was living in a homeless shelter, and only his therapists, teacher, and administrative staff knew. However, due to his mother's inability to wash his clothes, his hygiene was odious at best, and he was being mercilessly teased. Asking kids questions about birthdays or holidays—anything where money is usually spent is making an assumption, and can put the child in an embarrassing situation. It is better to not ask, and wait for the child to bring the subject up him or herself.

Another difficult time of the year is Mother's Day and Father's Day. Many public schools require kids to make cards in class for these holidays. What if your mother is dead and your father is in jail and you haven't seen him in ten years? Requiring students to make cards is insensitive in these days. Maybe back in the time of Ozzie and Harriet, but those days are long gone. As I said, I have stencils in my office, and my kids know they are there. Usually, if they want to make a card, they will bring it up. When these particular holidays approach, I may ask a kid, "Do you want to do any stencils today?" There is a calendar hanging above my desk, which the children can see, and they'll usually reply, "Hey, isn't Father's Day this weekend? Yeah, I want to make a card for my dad." Or they'll say nothing. However, by not shoving the holiday in their face, I'm not reminding them of what is missing in their life. Many of the kids will make cards for people in their life who they feel is like a mom or dad to them. Black children tend to have a lot of "play" moms, dads, sisters, and brothers. They'll point out a staff to me on campus and say, "Keisha's my play mom."

Anniversaries of when a person died can also be a problem for children. They tend to hang on to anniversaries of the death of a close relative for years. And as I mentioned, the period between

Thanksgiving and Christmas seems to have a higher incidence of deaths—at least among the population I've worked with. I always make a note in client files when a family member died, so that as that time of year approaches I can be a little more watchful for changes in mood.

Being sensitive to how these children's lives differ from the staff that works with them on a daily basis is necessary for understanding their behavior. When school is coming to a close, and teachers and other staff (and I'm completely guilty of this too) are happily talking about the vacations or plans they are looking forward to, one needs to be cognizant that the children around you are listening, and absorbing your words. But instead of feeling happy for you, they are resentful and angry. Although you are feeling happy and anticipatory, they are feeling anxious and tense. Your vacation may be a cruise, or trip to visit family, or hanging out with friends shopping and relaxing. But these kids are anticipating weeks of boredom at best, and tension, hostility, and physical violence at worst.

So when two weeks prior to a break the kids start acting just a little bit more defiant, and have to be restrained a little more often—do not worry—it's not that you have failed in your job. It's just a normal case of Spring Fever. When they come back from their break, everything will be status quo.

14

Can We Talk About S-E-X?

One of the problems with at-risk youth being on long breaks from school is that they are even more likely to have unprotected sex and get pregnant during the break than a regular teenager. A word of caution—if you are going to work with adolescents, you need to be comfortable talking about sex. In explicit detail. I do not mean to imply that the kids are going to be discussing it in explicit detail (although many do—again, like gang-banging, the ones who are boasting—are not doing it—the ones who are doing it, keep it on the down-low). However, it will be part of your job to teach your kids sex education.

So how come it is okay to teach sex education, but not religion in therapy? Is sex education therapeutic to your client? If they are already sexually active, yes. Is it in the best interest in the client, yes. Because their parents, schools, social workers, foster homes, and group homes can not be relied on to teach them about the facts of life, including sexually transmitted diseases and the repercussions of having sex with an emotionally unstable girl who might later decide to file rape charges. So if this chapter makes you uncomfortable—take a long hard think before you take a job with this population.

Unfortunately, sex education in schools today is apparently non-existent. I'm not certain what *is* taught in health class, but the only thing I've learned after all these years is that the kids I've worked with know next to nothing about the human body. While working at the group home, I decided to have a sex-education discussion for group one day, figuring that the chance to talk about sex with the boys would finally engage them. I kept my language as simplistic as possible. In fact, I avoided all clinical words like "vagina" or "scrotum" and talked to these boys like they were toddlers.

These boys, who had been sexually explicit and crude when I first started working with them in an attempt to embarrass me, suddenly were blushing and looking at the floor, the ceiling, anywhere but at me. (They eventually admitted that they were embarrassed, because by that time the therapeutic relationship had been established.) We first covered female anatomy, and I asked them, "How many holes does a woman have down there." The answer I got back? Two. When I asked which hole the baby comes out of, I learned the baby comes out of the hole the woman pees from. So we had lesson number one—the female anatomy.

Sadly enough, I was to learn that they did not know much more about their own genital area. They did not know what the function of their testicles was (the production of sperm and hormones). They did not know that they needed to examine their testicles periodically for changes to ensure that they did not have testicular cancer, which primarily occurs in younger men, and can start in adolescence. A typical response to this is, "I'm not doing that—that's gay!" To which I pointed out that they seemed to have no problem scratching their balls in public, so what was the big deal about handling them in the privacy of a shower?

On a more humorous note, when we were discussing the function of the prostate, one of the boys who had a job and had arrived late came in during the middle of the discussion. I told him, "We're having a sex education course. We're discussing prostates." He immediately shook his head, sat down and cringed, saying, "Oh, no, I'm not down with that. I ain't got one of those."

We went on to sexually transmitted diseases, and I asked how many STDs they could name. They were able to name about half of them, and were woefully ignorant about how you could get them. Most of them were under the impression that vaginal penetration was the only way to get an STD. When I graphically explained how a doctor would stick a long q-tip up their penis and swirl it around a few times to swab the inside of their penis if they ever showed symptoms of an STD, *that* got their attention. Six boys flinched, covering their groins. I have discovered the best way to make an impression with adolescents is with pictures. The old adage about a picture being worth a thousand words is definitely true when dealing with syphilis or gonorrhea.

Sexually Transmitted Diseases Surveillance 2007
Percent of Total Cases Among Teens 10–19 Years of Age

	Caucasian			Black			Hispanic		
Chlamydia	Total	Men	Women	Total	Men	Women	Total	Men	Women
All Cases	319,703	69,523	250,180	524,791	150,377	374,415	204,853	47,391	157,462
% 10-19	32.9%	17.1%	37.3%	39.6%	28.9%	43.9%	30.8%	22.1%	33.4%
Gonorrhea									
All Cases	69,767	26,207	43,559	250,245	125,006	125,238	30,680	14,733	15,947
% 10-19	23.1%	10.4%	30.7%	30.9%	21.1%	40.7%	25.8%	17.0%	33.8%
Syphilis									
All Cases	4,050	3,669	381	5,273	4,169	1,104	1,887	1,719	167
% 10-19	2.3%	1.3%	12.1%	9.1%	7.3%	16.2%	4.8%	3.8%	15.6%

Source: Centers for Disease Control and Prevention

According to the Center for Disease Control and Prevention, lower socio-economic groups are at higher risk for STDs. In 2007, almost half (48%) of all chlamydia cases occurred among blacks. Over two-thirds of the cases of gonorrhea was among blacks (70%). As the table above demonstrates, among overall cases, the percentage of cases among kids 10–19 is significant among all ethnic groups, especially among females. There is also a higher correlation between STDs and persons entering jails or juvenile correction facilities.

Special Education Schools are a hotbed of romances between students. It is a continuing soap opera of who is dating whom. With some students it is hard to keep track of their boyfriends or girlfriends because they change so frequently. Whenever an adolescent student starts at my school, I always advise them that if their goal is to return to public school, to stay away from the drama—especially the females who try to create more drama than a Hollywood producer.

One of my higher-functioning high school students was in a relationship with another student who was lower-functioning and a drug addict. Her entire family was probably doing drugs (I had her sister in counseling). Terrence and I kept discussing why he stayed in the relationship, and came to the conclusion that because his family life was so lonely, he would rather be in a dysfunctional relationship with a girl than no relationship at all. Well, they were sexually active, and it was consensual, but then the girl went crazy, and according to the story, her mother convinced her to press rape charges against my client. The charges were eventually dropped, but for months my client had to deal with that on top of his regular problems at home.

Whenever one of my male clients starts to show an interest in girls (and the age can range from 11 to never), I have "the talk" with them. Basically I ask them what they know about sex, STDs, birth control, protection against STDs, and then I tell them to assume the girl is always lying. If she says she is on birth control they still need to use a condom in case she is not actually on birth control, and also to protect against STDs. Unfortunately, when dealing with love, every adolescent says the same thing, "Oh, she would never do me like that."

Such was the case with Tyrone. Tyrone was always hanging on the girls around campus. We had more talks about sexual harassment and how he was setting himself up for some unstable girl to slap a lawsuit on him than I could count. His sexually inappropriate behavior came up in his IEP every time. He went through girl friends like dogs go through chew toys.

One day he announced he was with someone new. It was a girl in the high school building. My response was, "Whatever. It won't last. You don't stay with anyone long." A year later he was boasting about how they were still together. Of course, their relationship did not seem to be helping either of them. They were physically inappropriate on campus, and therefore both were getting in trouble. Neither were taking responsibility for their actions.

Then one day Tyrone did not want to come to my office. This was unusual. He told me, "I have to take the WRAT test." This is a standardized test, which I knew was not being administered that week, and even if it was, Tyrone would prefer counseling to a test. "I have to help the staff," he pleaded. I knew something was up, and grabbed him by his hoodie, and said, "You're coming with me. Tell your staff." I should point out that Tyrone is about a foot taller than me.

He came, reluctantly, and in my office we played cards together, but he kept losing, which was unusual. He was talkative, in a nervous sort of way, but his concentration was clearly elsewhere. I kept asking what was up with him, and he kept saying, "Nothing, Miss Jillian. Why?" Then he asked me when his next IEP was, and I reminded him that he had one about two months ago, which he knew very well. Tyrone confessed that they had held an emergency IEP the previous week.

I put my cards down, "Oh, really! Why was that?"

Tyrone started sliding down in his chair, "My parents wanted to kick me out of the house."

"What did you do?" I demanded.

By now, Tyrone was hiding behind the cards which he was holding up in front of his face. "I got my girlfriend pregnant," he mumbled.

For the next several minutes, Tyrone listened to me lecture him. We had been working together for four years. Hadn't I taught him about birth control? Hadn't I told him to use a condom, and never trust the girl was using birth control? Needless to say, I was disappointed in Tyrone, and he knew it, and that was why he was so reluctant to tell me. I think for him, telling me was worse than telling his parents. According to Tyrone, he thought his girlfriend was on birth control, but she stopped without telling him. They had used a condom, but it broke. Personally, I do not believe they used a condom, but if that's the case, I know Tyrone will confess eventually. He may have poor judgment, but he does not lie—at least not to me.

Sadly for Tyrone, he could have had a good future. Now he will be a father before he turns eighteen. I advised him to get a paternity test, because quite frankly, while there are fewer girls at my school, they tend to be more manipulative and unstable. I would hate to see a young man's future ruined because he wanted to do the "right thing," and it was not his child.

In a study by the Center for Disease Control and Prevention (2009) on preventing teen pregnancy, they found that there were 435,436 births to adolescents ages 15–19 in 2006. This represented a 3% increase compared to 2005. Hispanic girls 15–19 were more than twice as likely to give birth to a child than whites, while African-Americans were 50% more likely than a Caucasian to bear a child. Youth in foster care were also more likely to give birth to a child, while the least likely group to have a child was an Asian/Pacific Islander. Keep in mind, these statistics relate to births—not actual pregnancies, and do not take into account teens who have become pregnant and have had abortions or miscarriages.

What does all these mean in terms of the lives of the children? In terms of the teenage mothers, they are more likely to drop out of high school and to remain a single parent compared to girls who wait until age 20–21 to bear a child. As for the children of a teenage mother, they are more likely to perform poorly in school, be a victim of abuse, be placed in foster care, be incarcerated during adolescence or their early 20s, drop out of high school, also become a teenage mother, and be unemployed as an adult.

Then there is the flip side of the coin—when the parents do not want their daughters on birth control because of the misguided perception that this will lead to sex. Well, I'm here to tell you—*they're having sex anyway!* One of my clients was distraught after the summer break, because she had missed her period. She decided she was going to keep the baby, but as she became further and further along in her pregnancy, the mood swings due to the hormones caused her behavior to become worse than usual. Let me say this—teenagers are not meant to have babies. Their hormones are already raging out of control—add to that the additional hormones caused by pregnancy, and it is a prescription for suicidal ideation, depression, and out of control behavior—even in a "normal" adolescent.

Mary started talking about how when her baby was born she was going to throw it from a window and kill it. She was thinking of ways to cause herself to miscarry by exercising vigorously, or starving herself. Her life was ruined. She didn't want to be a mother at sixteen. Her life and dreams of college were over. Her boyfriend was a no-good for nothing who couldn't help her financially or emotionally.

After assessing Mary for whether she was really a danger to herself or her unborn child, Mary and I talked about her options. We discussed the possibility of adoption and the difference between closed and open adoptions. We talked about having her mother raise the baby, and finally we talked about abortion. Mary was adamantly against the latter.

Winter break came and went, and when I returned to school Mary requested to see me. She was crying and said she thought she was depressed and needed to go on anti-depressants. She hadn't been sleeping since the break and felt like she was "losing it." I asked her what had happened over the break, and in a nutshell learned she had aborted the fetus. I asked her if she wanted to talk about it, and she said no. I told her if she changed her mind, to let me know. I assessed her for suicidal ideation, and we made arrangements for her to see the psychiatrist and get immediate medication.

Fast forward to spring break and Mary sees me after the break, and looks at my calendar. Cognizant of the fact that there is a small child in the office next to mine, Mary says quietly, "I'm late." She points to a date on the calendar. "I was due there." I raise my eyebrows. She says quietly, "I had s-e-x."

I shake my head in disbelief. "You're not on birth control?"

Turns out Mary's mother won't let her get any birth control because she thinks if Mary has it she will be sexually active. And yes, Mary's mother knew that Mary had been pregnant the year before. I tell Mary that there are clinics that she can go to where she can get birth control, or at least condoms. Turns out that it was stress-related, and Mary got her period a week later. She was practically dancing down the school hallways with glee, and quite honestly, I couldn't blame her. Fast forward a year later, and Mary is now on birth control, has had a PAP smear, and is taking control of her medical life at age seventeen. As they say, better late than never.

On a completely different note, you may have to get sexually explicit with younger children to ascertain if sexual abuse or other inappropriate behavior has occurred. Again, it helps if you feel comfortable yourself discussing these matters, because children pick up on your anxiety and react to it. When I work with very young children, I will use whatever language they use. If they refer to their penis as a "pee-pee," then that is the term I will use. If they call it a "wangle," I'll use that term. If they simply point down to their crotch, I'll whisper, "penis?" in an encouraging tone to let them know it's okay to say the word. Girls almost invariably reply to their breasts as "boobies," and their vagina as "down there." While I expect adolescents to know the name of their vagina, with elementary school girls I use their terminology since most of them will not have heard the word "vagina" yet.

I had two male clients, both young (under ten) who were in the same class together. Mike had been sexually molested as a young boy. Steve had no abuse history to our knowledge. One day I was informed by their teacher that Mike told her that Steve had touched his penis while they were in the bathroom together. I pulled Mike out of class, and we went to my office. We played on the floor together as usual, and during the course of our play I asked him if anything was bothering him. He replied in the negative. I asked him if he was having any problems with any classmates. And he told me no. I said casually, "Oh, I was worried that you and Steve were having a problem."

It is best not to ask leading questions that a child can then give you the answer you want in order to please you. I did *not* ask, "I understand Steve touched you in the bathroom." What I learned from Mike was that while he was going to the bathroom, Steve attempted to show Mike the proper way to hold his penis while urinating. This fitted with Steve's profile, since he tended to be a "helper" in the class. There was nothing sexual about his behavior, although it was inappropriate. Mike, having been sexually abused, reacted more strongly than an average child might. We finished our session, and Mike was fine.

I then had a session with Steve, and it was in the same manner. We played casually, and during the course of our play, I asked him if he understood about respecting people's personal space—like when the students stood in line in school they were to keep arms length from each other and their hands behind their back. I then went on and explained that the same rules applied in the bathroom, and that we never touch another student, not even a hug, without their permission. I didn't go into explicit details, because I did not want to shame him, but he got the point. Steve was intelligent, and easily able to follow directions. I followed up with the teacher and Mike's parents, and by the next day Steve and Mike were close friends again and the matter apparently forgotten.

15

"I Think I'm Gay"

Hand in hand with discussions about sex, may be discussions about sexual orientation. Regardless of your own personal religious, political, or moral views, some of your clients will be gay. Some of your clients will have gender identity disorders. Some of them will simply be normal adolescents who experiment with the sexual continuum. Again, as a therapist it is your job to support, not to preach. To educate your client about sexual safety; not to convert. If the last chapter made you uncomfortable, this one may make you even more so. If it does, you may have some personal issues to work through before following this career path.

The reality is that when you throw adolescents into a forced living environment such as a group home, residential facility, or correctional facility—they will form relationships. Sexual ones. Does this mean they are homosexual? Not necessarily. Some are; whereas, some are simply starved for human physical contact and closeness. A superior of Mary Gilligan Wong, a former nun who left the convent and became a psychologist, said it best in Wong's memoir, "Look at it this way: you're locked up with a bunch of women day in and day out without any contact with men. Before you know it, you're having some stirrings toward someone of the same sex. It doesn't say anything at all about your basic sexual orientation—in fact, some would say you're a lot healthier if you *do* have some of these stirrings in a situation like this. At least you're feeling *something*, instead of letting your feelings dry up inside you!"

In the work I have done, females, rather than males, are more likely to form romantic relationships with same-sex partners, while males are more likely to be gay or have gender identity disorders. Nearly all the males I worked with who had gender identity disorders were sexually molested and/or physically abused as young children.

The females who questioned their sexual orientation tended to be normal adolescents going through their quest for identity. That is not to say I have not had gay female clients—but gay clients usually know from late childhood or early adolescence that they are gay.

Melissa came to me one day in session and was slightly agitated. We had been working together for over half a year, and had an excellent relationship. She had made a lot of progress in her behaviors and goals. Finally, she stammered out, "How do you know if you're gay?" I asked her why she was asking, and it turned out that there was a girl on campus that Melissa was obsessed with. She found her attractive, wanted to be like her, wanted to be with her. It should be noted that Melissa had a lot of self-esteem issues, and what I suspected was going on was that this was more a case of envy and a desire to emulate the other girl rather than feelings of attraction and love. However, it was also true that Melissa had never gone on a date with a male or been intimate with a male.

There were a lot of possibilities here. Melissa could be overwhelmed by the idea of being with a male due to her self-esteem, and a girl might seem safer. It's easier to be with someone intimately, when you've been friends with someone first. Given that Melissa was nearly seventeen and had never dated, she could actually be gay. The reasons for Melissa's attraction was not the issue. I was not House, with a problem to solve. I was a therapist, with a confused adolescent who had enough self-esteem issues already without being worried about being rejected for being gay.

I like to use analogies with my clients that make things understandable to them. So I took a piece of paper and drew the following, explaining to her this. I should state that this is my belief system, and others may disagree. But I've found this explanation tends to reassure kids and regardless of their true sexual orientation, they will find their own way eventually, no matter what is said to them by a therapist, priest, or politician. The only difference is whether they will live with their lifestyle in comfort or in shame.

I talked to Melissa about artistic ability, and how everyone is different. Artistic ability, like musical ability, or athletic ability, is something you are born with and it is on a continuum. Some people can pick up a pencil and draw anything that they see. They start drawing at an early age, and end up being famous artists. At the other end of the spectrum are people who can stare at a simple cartoon and are unable to recreate it to save their life. Most people fall somewhere in the middle. They have some ability to draw simple things, and that's about it. The point is you can't change that ability. If you are born with no musical ability, no matter how many music lessons you take, you will not become a concert pianist. Music lessons cannot change a person from being tone-deaf.

Sexual orientation is also on a continuum. Some people are born strictly heterosexual. Some people are born strictly homosexual. In the middle are people who are bisexual. On either side of the middle are people with homosexual leanings, or heterosexual leanings. I explained to Melissa how when people go through adolescence, their hormones increase, which can cause many things such as moodiness, tearfulness, anger, acne, and often this is when teens are prone to fall in love easily—because of raging hormones. I then let her know that it was not uncommon during this period for teens, particularly girls, to experiment sexually with the same sex. Did this mean she was gay? I did not know. Only time would tell. But her feelings were not abnormal; other girls went through the same thing, and whether she was straight or gay, didn't matter. What mattered was how she felt about it.

Melissa listened to everything I had to say, and ended up in the end replying that she didn't know if she was gay, but she did feel more normal and less scared than when her session started. "I thought there was something wrong with me." I assured her I was there for her to talk to if she wanted to in the future, and if she wanted me to recommend some readings to her I could.

Shawna, a bi-polar client, had multiple relationships—all negative and self-destructive with numerous boys. One day, when in a manic cycle, she announced in my office that she had a girlfriend, was gay, and had always been gay. As I stared at her, my eyebrows trying to crawl up into my scalp in disbelief, Shawna shouted, "What? Why are you staring at me like that?"

"*You're* gay?"

"Totally! Fuck men. I've always been about the girls."

Again I gave her an incredulous look, and then went into a lecture about how she needed to learn to recognize her manic phases—a talk we had had on numerous occasions. Shawna insisted she was gay, was not into men (despite having had more boyfriends than I could count). However, it should be noted that her last relationship had ended badly, with violence and a restraining order. In Shawna's case, I think that, coupled with her mania, caused her "normal" experimentation. As it was, a few weeks later she and her "soul mate" were broken up. Approximately a year later, Shawna was pregnant. When she told me, I said sarcastically to her, "I knew you weren't gay," which prompted her to break up into a fit of laughter.

Males, on the other hand, tend to deny being gay even if they could host their own television show on Bravo. Part of this is cultural. Since most of my clients are black or Hispanic—acceptance of homosexuality in both of these cultures is even lower than the population at large. I remember Guillermo, a young man in high school who was very sensitive, well-behaved, and popular with most of his peers. He did not come across as flamingly gay—just slightly effeminate. However, among his close friends on campus it was apparent he was gay, and they pressured him repeatedly to come out of the gay closet and just admit to who he was. However, during the entire eighteen months we worked together, he never brought it up, and I did not either since it would have violated the confidentiality of my other clients who discussed him with me. Additionally, his sexual orientation was not why he was in our school, it was due to other behavioral problems which interfered with his academic education.

While working in Watts, several of my black male students were flamingly gay, and after working with me for a year admitted to it, but it took that long for them to trust me with their secret. Interestingly, once they admitted to being gay, they became even more "gay" in session, complete with hand gestures, falsetto voices, and using the phrase "girl friend," when referring to me. Of note, once these student came out of the closet, they tended to have problems with violating physical boundaries with me—I was now "safe" because I was no longer a person of the opposite sex, so they could physically touch me, and I had to constantly remind them that I was still their therapist, and that hanging on me like a friend was not appropriate.

Perhaps the most difficult clients to work with, and the most tragic, are the students who have a gender identity disorder. The reason why I use the word "tragic" is again most of these children are black or Hispanic, and if homosexuality is misunderstood by lower

socio-economic parents, foster parents, or group homes in which these
kids live in, then gender identity disorder is even less understood.
Below is the DSM-IV summary for children and adolescents—again,
refer to the complete DSM for the entire criteria.

The DSM-IV Criteria for Gender Identity Disorder
(In Children)

A strong and persistent cross-gender identification (not merely a desire
for any perceived cultural advantages of being the other sex). In
children the disturbance in manifested by four (or more) of the
following:

1. Repeatedly stated desire to be, or insistence that he or she is,
   the other sex
2. In boys, preference for cross-dressing or simulating female
   attire; in girls, insistence on wearing only stereotypical
   masculine clothing
3. Strong and persistent preferences for cross-sex roles in
   make-believe play or persistent fantasies or being the other
   sex
4. Intense desire to participate in the stereotypical games and
   pastimes of the other sex
5. Strong preference for playmates of the other sex

In adolescents and adults, the disturbance is manifested by symptoms
such as a stated desire to be the other sex, frequent passing as the other
sex, desire to live or be treated as the other sex, or the conviction that
he or she has the typical feelings and reactions of the other sex.

   In the group home where I worked was a young man, Gerome,
whose favorite role models were all female entertainers. He would cut
out their pictures from magazines; however, there was nothing sexual
about this behavior, it was more that he idealized them and wanted to
be them. He walked around with a towel on his head pretending it was
his flowing long hair. While all children role play and gender play at
some time in their development, Gerome was sixteen at the time. He
also had makeup, which he insisted was for a drama class, but he wore
the makeup around the group home, and his make up was not

"theatrical" so much as feminine. Gerome had no male friends out side of school—all his friends were female.

Sadly for Gerome, all the boys in the group home tortured him. He was an emotional, nice young man. He loved all the female staff at the home. He was affectionate and loved to hug the women. Again, none of this was sexual—more in a needy, nurturing way, as though he was seeking the maternal love he never received from his abusive family. The boys tortured Gerome because they called him a faggot and gay. I seriously doubt Gerome was gay, because he did not display gay qualities and showed no interest in men sexually. Gerome's problems seemed more rooted in a hatred of his own sexuality. I worked with Gerome for two years, and while he discussed all the right things—the problems he had with peers in the home, problems at school, treatment goals—he was unable to open up to me about the sexual atrocities he had endured as a child or his gender identity issues.

At my school, I had a young man on my case load very briefly who reminded me of Gerome. Wayne was possibly gay, or possibly had a gender identity disorder. It was difficult to tell, and he did not stay at the school long enough to discern the difference. He was very effeminate, and when I asked Wayne what he wanted to be when he grew up he said "Britney Spears." Wayne was very into fashion he would tell me. He spoke in a high, falsetto voice, had gay mannerisms, and preferred the company of girls—but as friends, not sexually. Of course, Wayne had a host of other problems, including slight mental retardation, and possible exposure to drugs in utero. Sometimes, when working with multiple diagnoses—things aren't always as black and white as the DSM-IV would have us believe. Either way, Wayne's path was going to be a difficult one in terms of acceptance from his peers, no matter what label was put on him. My job, should he have remained at our school, would have been to help him cope with that lack of acceptance.

16

Sticks and Stones. . . .

There's that old childhood taunt about sticks and stones may break my bones, but words can never hurt me. Unfortunately, words hurt sometimes even more so, because they do not leave physical scars. Emotional abuse by parents can be terribly damaging to a child's self-esteem and sense of self worth. I have had children tell me how their parents make it clear that they were "an accident," or that they want to "send them away." Children who were born to mothers who were teenagers themselves when they became pregnant, often remind their child that if they had it to do over again, they would have given their child up for adoption, or had an abortion. When children hear these things over and over, they begin to feel less than human. They feel like a damaged object that their parents want to discard—but can't.

Special needs children get enough teasing in school from their peers on a daily basis. They get called racial epithets, names based on their looks, or based on their low cognitive functioning. Usually, the kids doing the name calling have their own self-esteem issues that they are struggling with, but this does not excuse their behavior.

It is one thing for these students to have to endure name-calling and teasing from their own peers, who are in the same environment as themselves due to emotional disturbances. It is quite another for a child to endure teasing, or unprofessional remarks, from a staff member.

As mentioned earlier, these students have a difficult enough time forming attachments to staff due to the constant changing of staff in their multiple environments (schools, foster placements, group homes). If they hear verbal abuse from a staff—how much trust can they place in the adults who are theoretically there to care and protect them?

One of the most unprofessional gaffes I ever heard a co-worker say was when I was walking a blind student to my office (her cane was left at home), and she was walking slowly, leaning against me. A therapist came towards us with a student, and said to his student, loud enough for my client to hear, "See, now *that* girl is probably autistic." Needless to say, my student was upset because she was not autistic, but aside from that, even if she was, why should this therapist point it out to his student? Autistic students are not deaf and they have feelings. I corrected the therapist, saying loudly, "She is *not* autistic, she is blind," at which point he apologized. However, that was not the point. How would this therapist felt if I had been walking with a student and pointed to the child on his caseload and made a comment such as, "See, now that boy is probably a bully?" For one therapist to comment on another therapist's client is completely unprofessional, and has no therapeutic purpose in the context of the work with his own client.

One day when I was walking a student who was particularly large back to class, I overheard an aide who was sitting at a table, say, "Wow, I can't believe how much weight Alan's gained since I've been here." My student was an AWOL risk, otherwise I would have stopped him and confronted the aide on her disrespectful comment then and there and demanded that she apologize to my client. However, my student was walking rapidly, and I could tell by the way he was clenching his fists that he overheard her remark and was upset. I caught up to him and said in an upbeat tone, "Good job ignoring others," and patted him on the back. The ironic thing was the aide who made this remark was at least 50 pounds overweight herself.

Whenever I go to a teacher to discuss a child, I always ask the teacher to step out in the hallway and we close the door so that the student cannot overhear the conversation. Usually, because I have multiple students in every class, my client does not know who is being discussed (although the more paranoid ones always assume it is them!). However, there is one teacher who never shuts the door, so I normally take several steps down the hallway for privacy. This teacher refuses to budge from his open door, and will discuss the student loudly, so that the student can hear every word being said. He will speak in a derogatory and shameful tone, making certain that *all* the students in the class hear him.

One of the very first days I was working at the school in the Valley, I was walking across the physical education area, when a boy, who had limited gross motor skills and was slightly overweight, was trying to ride around the quad on a scooter. He fell, and two of the staff sitting at a table, bent over laughing so hard they were unable to control themselves. These staff were not simple child-care workers who often are working on their bachelor degrees. They were theoretically staff with a Masters level education. I immediately walked over to them, and said, "These kids have enough problems with self-esteem without you laughing at them!" Their response was to say, "Lighten up! It was funny!" Needless to say, the child did not attempt to ride the scooter again, and I do not think it was a result of his fall.

When working with a child and talking to them about their behaviors, talking behind their backs on a playground, or at a group home or residential facility—consider this one thing—*would I want someone making this same comment to _my_ child's face?* If the answer is "no," then you are probably being inappropriate, and need to look at why you feel a need to put someone down who is younger than you when you are being paid to help them. Is it countertransference? Do you have job burnout? Does it make you feel superior? Or is this the wrong profession for you?

There is a right way and a wrong way to confront a child on their behaviors while working in an open environment. I have a reputation at my school for being one of the most strict and mean therapists. My style is, to say the least, confrontational. However, I confront with love. And I do it without shaming students or calling them names. There is a difference between being therapeutic, and recreating the environment of abuse these children already experience at home.

I was assigned Todd after he had been at the school for over a year. How he had slipped through the cracks and never received counseling is beyond me. Todd was large for his age, overweight, an instigator at best, a bully at worst, but academically bright and obviously carrying deep pain within his soul. He loved to read, and had a love of words, so in my office we did word searches, boggle, and scrabble. Todd made it his personal goal to beat me at scrabble or boggle. He would stay in session over his allotted time in order to finish the game.

The one thing Todd did not do in session was talk. I would hear from staff that he had a bus chart, was on LOP, had an incident report, et cetera. I would bring these things up in session, and he would shut down, and not answer my questions. Approximately three weeks before a break (is this starting to sound familiar yet?) Todd refused to come to session with me. He claimed it was a waste of time because all we did was play games. He eventually came, but refused to answer my questions. He was polite, but distant. He said everything at home was great, he was getting along great with his parents, everything was fine. I thanked him for turning things around and coming to my office and returned him to class.

Meanwhile, Todd was getting more and more out of control in class. He was destroying property and physically assaulting his classmates. I witnessed him throwing items at peers who were literally one-third his size. Every other word out of his mouth was "fuck" or "bitch." In therapy, when I attempted to discuss his behavior, he denied any responsibility, calling me a bitch, and covering his hands over his ears, saying, "I'm not listening," while singing "la, la, la" at the top of his voice. However, the entire time he was behaving defiantly in my office, he did not make eye contact with me, leading me to believe it was not me he was angry at, but someone or something else.

So I called his parents to see if things at home were okay. They claimed everything was status quo; that Todd was still being Todd. His parents had an AB3632 in place already (residential referral), and I told his father that, quite frankly, if Todd's behavior did not improve, at his next IEP my recommendation would be removal from our school.

The following week, Todd was furious at me for calling his family and refused (for the second consecutive week) to come to counseling. I knew Todd had a semi-good relationship with one of the aides in the class, named Brian. I asked Todd to step outside, with Brian. This is referred to as making the best use of your resources. If Todd wouldn't respond to me, maybe he would respond to Brian.

To put it bluntly, I got into Todd's face. I told him that he needed to cut the crap, and make a choice. He had two options. One, to start coming to counseling and tell me what the fuck was going on with him to cause his behavior to deteriorate so badly, or two, come October in his IEP, I was recommending he leave the school. Then I spelled out for him that there was already an AB3632 in place, and that I had worked at the school for five years and the entire IEP team trusted my opinion and if I said he needed to go—he would go. I said he had one week to think about it, and to make his decision. The choice was his, but he had better be prepared to accept the consequences.

During this talk, Todd kept trying to interrupt me and tell me he wasn't doing anything, it was the other kids' fault, and so on. Every time he interrupted me I held up my hand and said, "Stop talking." I would turn to the aide and ask a question. "Brian, has Todd been instigating his peers?" The aide would confirm this. "Brian, has Todd been physically inappropriate?" And so forth. With a witness that Todd had a modicum of respect for, Todd had no comeback. I sent Todd back into the class, and then said to the aide, "See what you can do." I heard Todd say loudly as he went back inside his class, "I'm gonna get that fucking bitch fired!"

Two hours later, I was taking a student back to class, and Todd's class was coming inside from PE. Todd tugged on my sleeve and said, "I want to come to counseling." To be perfectly honest, I was shocked. But I didn't let him know that. I just nodded, and continued on with my other student.

After lunch, I went to Todd's class and asked if he was ready to see me. He came to my office, and I turned the tables on him. I started the session by saying, "I want to apologize to you. Apparently, when we first met, I failed to tell you the one rule of my office. You are allowed to express yourself anyway you want. If you want to swear, that is perfectly all right with me. However, I expect you to respect me, and to respect my belongings. Now last week, by behaving the way you did, you failed to respect me. So I'm going to assume it was my fault for not telling you my rule. You have always been respectful to me, and I think otherwise you would not have sworn at me the way you did."

Todd sat there, staring. He of course apologized for swearing at me, and I accepted his apology.

I then asked him, "Do you think I like you?"

Todd shrugged, "I don't think you liked me last week."

"Let me clarify something. I may not have liked your behavior last week, but that does not mean that I still don't like you. I recognize that you are a smart, funny, personable young man with a good sense of humor. When you misbehave—I may not like how you behave in that moment, but that does not change how I feel about you overall." Todd seemed to struggle with this concept. Then I asked him, "Do you think I care about you?"

"I don't know."

"Let me put it this way," I said. "If I didn't care about you—if I didn't think you had any potential to better yourself, I would not have bothered to call your family to see what was going on at home. I could have said—oh well, Todd's acting up. So what. Let him get out of control and get taken away by the cops and put into the Halls. So what? Who cares? I don't. But I do care, and that's why I called to make sure nothing had happened, like somebody had died that you were close to, to cause you to suddenly start acting out like this."

The rest of the session Todd went on to tell me things about his family and his feelings that he had never revealed in the ten months that we had been working together. There are many other ways I could have handled this situation. When I observed Todd's behavior in class I could have shamed him in front of his peers. I could have called Todd names—fat, lazy, a loser, a bully. I could have had my conversation in front of his peers instead of the privacy of a hallway. I could have demanded when he finally did come to my office that he apologize to me for his behavior. I may have jumped all over his case, and not let his behavior get out of control while I was talking to him, but I spoke to him as the teenager he is—expecting him to make choices and accept the consequences for his actions. By treating him with respect, I was able to convince him that I was on his side.

There are many ways to deal with children's issues in a *constructive* way, rather than a *destructive* way. If a child is overweight, talk to them about nutrition. Do not make fun of their weight, call them names, or laugh at them when they attempt to exercise on the PE grounds. Many of these children have poor gross motor skills and are unable to play games such as badminton, volleyball, or baseball. Even Frisbee is difficult if you have cerebral palsy, downs syndrome, or Autism and have problems with hand to eye coordination. Simple catch can be challenging. Staff should be applauding any student's efforts to better themselves, not making derogatory remarks. Even if these remarks are made out of the child's hearing—their peers will hear the remark and pass it on.

The teacher I mentioned earlier who had a continual habit of shaming his students in his class, or when he was discussing them with me outside the classroom, was assaulted by several students one day. I do not know the circumstances of what happened, but I do know this.

I have never known a teacher who treated their students with respect to be physically attacked.

If you work in a field with children, think about this—the next time your child comes home and complains that a teacher at school made fun of them, ask yourself this question—*Have I ever committed the same crime in the capacity of my job?*

17

Pervasive Developmental Disorders

When I first started working at the school in Watts, I worked with numerous children with Autism. At my current school in the Valley, I have worked with a few autistic kids, but primarily I have had more kids with Asperger's Disorder. Everyone is always asking me, what is Asperger's? Asperger's, like Autism, is classified as a pervasive developmental disorder. Both are diagnosed in early childhood, usually after the age of two or three. The best way to understand the difference between the two, is to show case examples. But for educational purposes, I will also include here the DSM-IV criteria. However, until you actually meet a student with both disorders, seeing the criteria of both can be confusing.

Asperger's Disorder
DSM-IV Diagnostic Criteria

A. Qualitative impairment in social interaction, as manifested by <u>at least two</u> of the following:

1. Marked impairment in the use of multiple nonverbal behaviors such as eye-to-eye gaze, facial expression, body postures, and gestures to regulate social interaction
2. Failure to develop peer relationships appropriate to developmental level
3. A lack of spontaneous seeking to share enjoyment, interests, or achievements with other people
4. Lack of social or emotional reciprocity

B. Restricted repetitive and stereotyped patterns of behavior, interests, and activities, as manifested by <u>at least one</u> of the following:

1. Encompassing preoccupation with one or more stereotyped and restricted patterns of interest that is abnormal either in intensity or focus
2. Apparently inflexible adherence to specific, non-functional routines or rituals
3. Stereotyped and repetitive motor mannerisms (e.g., hand flapping)
4. Persistent preoccupation with parts of objects

C. This disturbance causes <u>clinically significant impairment</u> in social, occupational, or other important areas of functioning.
D. There is <u>no</u> clinically significant general delay in language
E. There is <u>no</u> clinically significant delay in cognitive development or in the development of age-appropriate self-help skills, adaptive behavior (other than in social interactions).

The children I have worked with who have Asperger's tend to be male, white, and as indicated by the DSM-IV, have no delay in cognitive development. In fact, the opposite tends to be true, and most of the kids who I have worked with who have had Asperger's have had above-average intelligence. Unfortunately, this often makes it difficult for their parents to accept that their child has a problem. Autism, which tends to be more extreme in overt behavioral problems is usually diagnosed early on when speech is delayed or lost. Because children with Asperger's can be precocious when very young, parents generally do not take their children to the doctor to find out what is wrong with them when the other behaviors, which are more subtle, begin to manifest themselves. It is usually not until these children enter school, where social interactions are more demanding, that the behavioral problems arise.

Additionally, in my current school where most of the students overall tend to have learning disabilities, cognitive impairments, mental retardation, or other problems resulting in their performing below their grade level—this causes children with Asperger's who have above-average intelligence to have an even more difficult time fitting in. Their poor social skills make them an easy target for teasing, and their "nerdiness" is an additional factor in making them a social outcast.

Children with Autism and Asperger's tend to both be either hyper- or hyposensitive to touch and/or sound. Therefore a slight noise can overwhelm them, or they are insensitive to noise and speak in a loud voice, unable to regulate their tone because to them everything sounds the same. Hair-brushing, teeth-brushing, washing, dental visits, doctor visits—all these activities involving touch become torture devices for kids with Asperger's (autistic kids often run the opposite spectrum and hit themselves and engage in self-injurious behavior because they feel nothing—it is self-stimulating).

I had a child with Asperger's who was insanely loud—I always saw him first thing in the morning, before the other therapists arrived so he would not disturb them. I spent an entire year with him lowering my hand to indicate he was to modulate his volume. Because kids with Asperger's tend to be rigid and stick to routines, as a result of my always seeing him first thing in the morning, he became anxious if I failed to see him at that time, and I eventually realized I had to mix up his schedule to get him comfortable with change in routine. Another child with Asperger's was hypersensitive to noise, and whenever the children in the other counseling offices were loud, he would cover his ears, cringing as though in pain. He was also sensitive to smell, and would complain about smells that I could not even detect.

Children with Asperger's tend to be obsessive with strange hobbies—usually animals, some obscure branch of science, or other things that relate to facts and figures. I had one child with Asperger's who drew dragons week after week. Or any animal with wings or fire. While children normally go through developmental phases where they become fixated on some sort of play, they typically attach and move on. Girls play with Barbies obsessively, then outgrow it and get into makeup and boys. Boys play with tonka trucks, then T-ball, then basketball. The difference between normal development and Asperger's is the pervasiveness of the obsession, and the level of agitation when unable to engage with the obsession. If a small girl can't play with her Barbies because they were left at home and not brought in the car, she can be distracted with something else. Kids with Autism and Asperger's tend to have obsessions that last longer and are more intensified.

Ben was added to my caseload when he was in the sixth grade and we would work together for two years. He was academically bright, but unmotivated. He had poor self-esteem and suffered from depression as a result of not fitting in. Ben was a voracious reader who spent more time reading than socializing with peers. Unfortunately, while Ben had a lot of great qualities—he had a wonderful sense of humor, and could make great puns or jokes with a play on words due to his strong vocabulary, his lack of social reciprocity made it impossible for the other students to get to know these positive qualities about him. A typical exchange with Ben would go something like the following:

"So, Ben, how was your weekend?"

"Great! My sister and I went to Disneyland with my grandparents! We rode the Pirates of the Caribbean three times. Did you know that pirates date back to the early twelve hundreds, and originally they weren't really pirates, they were just traders, and they got the reputation for being pirates because sometimes the countries they would go to wouldn't let them land and they would need food and water and have to use their weapons to force the people to let them land so that's how the term pirate got started. Oh! And did you know that the oldest ship ever recovered with actual gold was off the coast of New Guina in 1836—not the date of the ship but the date they found the oldest ship" (This sort of tangential monologue would continue until Ben literally ran out of breath or I stopped him).

"Okay, so it sounds like you had fun at Disneyland with your grandparents. Remember Ben, what I told you about how come the kids at school might not like being around you?" (Here he would shut down for a minute.)

"It's not my fault! They're all stupid! They can barely read!" He crosses his arms protectively in front of him, looking angry.

"Well, maybe if you gave them a chance to talk, you might find out that you have something in common to talk about together. I asked you how your weekend was. What do you think an appropriate response to me would have been after that?"

Ben thinks long and hard, then shouts, "How was your weekend!"

I smile, "It was very nice, thank you for asking." I don't provide any information to see if he will ask me what I did.

"Oh, and I read a book this weekend. Did you know that there are 57 different types of sand in North America? Some of the types of sand you have to dig down–"

I cut Ben off, "Okay, remember what we talked about. Are we having a conversation right now, Ben? Or are you having a monologue?"

He says nothing, staring at the ground.

I say gently, "How do you feel when the teachers get mad at you for misbehaving, and they starting lecturing at you without letting you speak. They just go on and on, talking and talking. You shut them out, right?"

He nods, giving me a reluctant shrug.

"Well, when you start turning into 'Mr. Professor' when talking to the kids, and you don't let them speak, how do you think they feel?"

He looks defiant. "But I'm just trying to educate them!"

"I understand that, but it's the teacher's job to educate them. It's your job to be their friend. Do you want friends, or students?"

Ben and I worked over our years together on understanding why he relied on all his scientific quotes and data when communicating with people. Sadly enough, Ben was in social skills classes outside of school—although you would never know it. He did have excellent eye contact, but aside from that, all his other behaviors were completely lacking in social etiquette. As I have mentioned in my chapters on setting boundaries, I rarely bribe my students. However, at our school therapists are allowed to make contracts with their students. Ben was one of only four kids (one other client was also Asperger's), where I made a contract. It seemed to be the only way to get my point across with regard to his lack of communication skills.

I made an agreement with Ben that for four weeks, when he and I were together I would give him a visual warning when he started to "lecture" rather than have a conversation. He would get only one warning, and he had to correct himself on his own with just that one warning in order to earn his contract. For the visual cue, I had a plastic jar filled with marbles on my desk, and whenever he started to go off on one of his tangents, I would simply turn the jar upside down. Surprisingly, Ben was able to stop himself every time he saw the jar turned, think back to what the original *conversation* had been about, and get back on track. He earned the reward in four weeks and was very proud of himself for achieving this goal. At the end of the four weeks, I told him, "Think how much easier things would be for you in class if you could do this in your head without a visual cue. Then all your classmates would get to see what a great kid you are."

What Ben and I decided over our time together (he was thirteen when I stopped working with him), was that he used his scientific jargon to keep people emotionally at a distance, and also, because of his low self-esteem, it made him feel good about himself to be able to show how much he did know. Because the reality was, in our school all he ever heard was what he was doing wrong—which was a lot. When I had to give Ben a Devereux (a standardized test for students) prior to an IEP, he came up as seriously depressed, low self-esteem, isolated from his peers—basically all the red flags of a young man who needed serious help.

Below is the DSM-IV criteria for autistic disorder. The two primary categories in italics represent those that are the same for Asperger's Disorder for comparison purposes. As you can see, the primary difference between the two is the addition of a delay or impairment in speech, as well as the deletion of the criteria for normal cognitive development. That is not to say autistic children cannot have normal cognitive development, but most do not. There can also be, at least in the SED and foster care setting, mild to severe mental retardation. However, autistic children with severe, or even higher levels of moderate mental retardation do not receive counseling, since cognitively they cannot benefit or participate in such an activity.

Autistic Disorder
DSM-IV Diagnostic Criteria

1. *Qualitative impairment in social interaction, as manifested by <u>at least two</u> of the following:*

    a. *Marked impairment in the use of multiple nonverbal behaviors such as eye-to-eye gaze, facial expression, body postures, and gestures to regulate social interaction*

    b. *Failure to develop peer relationships appropriate to developmental level*

    c. *A lack of spontaneous seeking to share enjoyment, interests, or achievements with other people*

   d. *Lack of social or emotional reciprocity*

2. Qualitative impairments in communication as manifested by at least one of the following:

   a. Delay in, or total lack of, the development of spoken language

   b. An individual with adequate speech, marked impairment in the ability to initiate or sustain a conversation with others

   c. Stereotyped and repetitive use of language or idiosyncratic language

   d. Lack of varied, spontaneous make-believe play or social imitative play appropriate to developmental level

3. *Restricted repetitive and stereotyped patterns of behavior, interests, and activities, as manifested by at least one of the following:*

   a. *Encompassing preoccupation with one or more stereotyped and restricted patterns of interest that is abnormal either in intensity or focus*

   b. *Apparently inflexible adherence to specific, non-functional routines or rituals*

   c. *Stereotyped and repetitive motor mannerisms (e.g., hand flapping)*

   d. *Persistent preoccupation with parts of objects*

4. Delays of abnormal functioning in at least one of the following areas, with onset prior to age 3 years: (1) social interaction, (2) language as used in social communication, or (3) symbolic or imaginative play.

   The autistic child, Jose, that I worked with who when I initially met with him played with the playschool castle, had such repetitive and idiosyncratic speech patterns. While his vocabulary was excellent and he spoke succinctly and clearly, he communicated by repeating what he had heard others say. For example, when we first met and he was playing with the castle, and a character fell and was "injured," my response to this was, "Oh, dear. Does he have to go to a doctor?"

Jose's response was to reply something along the lines of "come to UCLA Medical Center, the Best Care in the World."

Clearly, Jose was repeating a television or radio commercial he had heard. While his answer made sense in the context of the question I had asked, it was as though he were unable to answer the question by choosing his own words. Just as he would later adopt the phrase, "The alligator's gonna bite," as a greeting to get my attention, he had difficulty expressing himself using his own words. His vocabulary was age-appropriate; his way of expressing himself was not.

Another autistic child who I worked with, who was much lower-functioning, was the complete opposite of Jose. While Kevon was able to read and do math (significantly below grade level), he was able to express himself excellently in terms of emotionality. He could tell you how he was feeling, why he was feeling what he was feeling, and had a concept of the past, present and future. Unfortunately, Kevon's speech was so impaired that he was virtually unintelligible, and I would spend half the session asking him to repeat himself or to slow down. Kevon would get excited (and start jumping and flapping like a bird), and the more excited he became the less intelligible he would become. His fine motor skills were the same. So while Kevon was capable of writing, spelling, and doing math, no one could read his work after it was completed because it represented the work of a three year old child. His numbers and letters literally ran into each other. At fourteen, he was unable to color a picture at the level above a toddler. He had zero spatial-perception skills or hand-eye coordination. In other words, the prognosis for Kevon was poor. Despite his cognitive abilities, he would never be able to get a job doing manual work in a factory, because his gross motor skills were so impaired.

Often, behaviors of autistic children, and occasionally children with Asperger's, can mimic psychosis. Because autistic kids tend to repeat things they have heard (echolalia), particularly movies or video games (which leads to the question of what their parents are letting them do at home since this is reinforcing their poor social skills), you can occasionally see a child suddenly start to "re-enact" scenes from movies, that taken out of context makes no sense and appears to be disorganized and incoherent behavior typically associated with a psychotic break. The above mentioned student, Kevon, was allowed to watch R-rated movies which were completely inappropriate for him, and when he became out of control at school, he would get up in the middle of the class and start screaming scenes from movies—making stabbing and killing gestures, and he would appear either suicidal or homicidal.

I had another student, a high-functioning Asperger's adolescent, whose diagnosis was not in his IEP. Harvey was in the ninth or tenth grade when he was assigned to me. One day the teacher called me, upset, for a consultation. She was worried because Harvey was talking and laughing to himself, and she could see no apparent trigger for his laughter. Miss Jones was worried that Harvey was schizophrenic. I told her that I had not seen any signs of this, but that he might be high-functioning Asperger's. He had poor social skills, and had no friends in his class as a result. I told her the next time he engaged in this behavior to page me, so I could come and observe him, but of course, like Murphy's Law, these behaviors never occur when the therapist is on campus or available.

In session I asked Harvey directly about his behavior. It turned out that Harvey, who's particular obsession was a television sitcom, would run the sitcoms through his head when he was done with his work, and that was why he was laughing. He was more or less watching television shows in his head. I knew for a fact that Harvey had the episodes memorized. He tended to communicate in session by giving me line-by-line dialogue of an episode. If I asked him why he liked an episode, he wouldn't say, "I liked it because it was funny the way the character handled the situation." Instead, he would say, "I liked it because character A said this, character B said this, then character A did this –" and he would literally tell the whole story. He was unable to extrapolate the plot or extrapolate his feelings. He often did these television shows in class, which was why his peers considered him weird. One, the shows he watched were those watched by children probably four years younger than him, and two, his inability to express himself as his peers did made him stand out.

I went back to his teacher and explained that there was nothing to worry about—or at least, nothing too severe. I explained to her that when he had completed his work, I had instructed Harvey to request additional work, or to ask to help out staff in some task rather than sit and do nothing. Like most Asperger's children, Harvey was bright academically and eventually graduated and passed his high school exit exam. His social skills continued to be his weakest area, but he was trainable. While that word sounds harsh, it's accurate. If I told Harvey, "You never say 'hi' to me when you see me on campus," he would start to do so the next day. But he would never take it to the next level. He did exactly what you told him to do, and not one iota more. Initiative was not his strength; but he was excellent at following directions.

Because children with Autism and Asperger's do have such ingrained obsessions with a particular toy, pastime, or activity, I tend to use my therapy time with them to take them outside their comfort zone, and break them of this obsessive behavior. For some of these children, this behavior becomes what is termed "self-stimming," in that it provides for them the stimulation that others achieve through social interactions. In addition, as I mentioned with Ben, the young man who loved to spout off scientific facts, their obsessive behavior is usually something they are good at. If I allow them to engage in this behavior in my office, I'm enabling their disorder. How will they grow if they do not try other things? It would be the same as if I allowed an alcoholic to drink during our sessions. These students are in the school to learn to become self-reliant and to stop behaviors that are going to impede their progress when they transition from school into the adult world.

I had one autistic student who was obsessed with puzzles. He could build a miniature 100-piece puzzle for adults easily. If he was left in a puzzle store, he would have built puzzles all day without stopping to eat. I made a deal with him that if he did a different activity every other week, on the alternating weeks he could build a puzzle. The activity had to be a different one—he couldn't play Connect Four every other week. The reason for this is often a preferred activity comes from being associated with someone or something pleasurable from the past, and by allowing the child to play the same game or toy the opposite week, they can simply switch their obsession to the other object.

I also used the removal of the obsessive object as a consequence of bad behavior. If the child misbehaved in session, the following week they would lose the privilege of their preferred activity. If they were really bad (physically aggressive, spitting, destruction of property), they lost the privilege for a month. This rarely happened— maybe once a year. But with such a low-functioning population, the threat of a consequence works better than a bribe. All I had to do with Kevon was say, "Calm down, or no Chutes and Ladders for a month," and he would immediately sit down, take a deep breath, and apologize.

One of the most frequent questions I get asked by teachers is about the difference between Autism and Asperger's. As a clinician, the differential diagnosis can sometimes be difficult to make because of the range. Just as depression can range from people who feel blue all the time but still function everyday, to people who stay in bed and want to commit suicide, pervasive developmental disorders also run on a spectrum. A child who is high-functioning autistic can look similar to a low-functioning child with Asperger's. How then, do you tell the difference? Below is a chart, with some simplified characteristics showing the range. Bear in mind, that there are ranges in between for mid-functioning on both ends of the spectrum.

Differential Characteristics of Autism and Asperger's Spectrum

Low-Functioning Autism	High-Functioning Autism	Low-Functioning Asperger's	High-Functioning Asperger's
Severe mental retardation	No mental retardation (possible cog delay)	Adequate cognitive functioning	Above-average cognitive functioning
No speech or speech w/echolalia	Good speech	Speech is tangential	Speech demonstrates poor social skills
No eye contact	Eye contact (may need prompting)	Adequate eye contact (may need prompting)	Adequate eye contact
Severe self-stimulatory behavior	Mild self-stimulatory behaviors	No self-stimulatory behaviors	No self-stimulatory behaviors
Severe obsessive behaviors	Moderate obsessive behaviors	Mild to moderate obsessive behaviors	Mild obsessive behaviors
No imaginative play	Imaginative play w/o social interactions	May lack imaginative play, or w/o peers	May lack imaginative play, or w/o peers
All behaviors isolated; no peer interactions	Peer interactions cause conflicts	Peer interactions are very poor	Peer interactions are fair

This is one of the most challenging populations to work with, and everyone has to make the decision about whether it is the right population for them. Personally, I do not work with autistic kids anymore. When I was at Watts, the autistic population represented the most clinically challenged students at the school, and approximately 15% of my caseload was autistic. However, since my current school has so many clients who are more clinically disturbed—schizophrenic, bi-polar, dual and multiple diagnoses—that dealing with Autism as well would be overload.

However, the Asperger's population I enjoy immensely. Some might say they are more difficult because their high intelligence can make them more manipulative. And as mentioned, it is easier for the parents to be in denial that their child has a disorder at all. When your child jumps up and down, flaps their hand, bites themselves repeatedly, and slaps themselves in the face, no parent is going to dispute their child is in need of special services and seek treatment.

One final note. No matter how blank an affect the child's face may appear, never be in doubt of this. Autistic kids have feelings. Autistic students are not deaf. They hear the teasing, and because of their obsessive behaviors, they are easy targets for taunting. I worked with one high-functioning autistic boy in public school whose obsession was Pokemon. The kids at his school knew this, and whenever they passed him they would shout, "Pokemon!" to get a reaction out of him. It worked like a charm, and he would start crying. These kids often look like they are catatonic and in a vegetative state with their rocking and repetitive noises. Sometimes, they are simply trying to cope with their emotional pain and disconnection from the world.

18

Psychotic Disorders

While there are other types of psychotic disorders, the primary one I have encountered while working with hard-to-reach adolescents is schizophrenia. So for simplicity's sake, I will limit this chapter to this one disorder, rather than focusing on other psychotic disorders. Chances are if you can work with schizophrenia, you can handle any other psychotic disorder you may encounter.

The DSM-IV characterizes schizophrenia as having two or more of the following characteristics for a significant period of time: 1) delusions, 2) hallucinations, 3) disorganized speech, 4) grossly disorganized or catatonic behavior, 5) negative symptoms, i.e., affective flattening, alogia, or avolition. There must be significant impairment in either social or occupational functioning, and there must be continuous signs of the disturbance for at least six months. This is a somewhat abbreviated definition—there are many caveats and exceptions, and to see the complete criteria I suggest referencing the DSM-IV. However, this is the gist of the disorder. There are also several subtypes of schizophrenia, such as paranoid type, disorganized type, et cetera.

I have worked with several students with schizophrenia, and the one thing I have come to realize is this—there is little you can do in therapy except to help them cope with their symptoms. Schizophrenia, like ADHD or bi-polar disorder, is a result of an imbalance of chemistry in the brain. The symptoms of the disorder cannot be corrected through behavior modification. All you can do is manage the symptoms. While this may be frustrating for therapists, teachers, or childcare workers—*you* do not have to live with the disorder for the rest of your life—these kids do. One thing I can say about my job—it makes me grateful for the little things, like having proper brain chemistry.

Unfortunately, most teachers are not trained on how to deal with psychotic behavior. The concept of delusions is foreign. We've all heard the expression, "He's delusional." However, this expression is often misused by the public. A person can say, "I'm going to be the best Hip-Hop Artist in the world—you'll see. In your face!" Is he delusional? Not if he actually can sing, and writes hip-hop lyrics. Someone who does that as a hobby has a chance of becoming a famous artist. Is he being *grandiose*—yes. Unfortunately, lay people often use the word delusional when what they mean is grandiosity—an expression of self-inflated sense of importance. However, it is based in reality and there exists a possibility of the spoken action being possible.

The stereotypical delusion is of the man who insists he's Napoleon, and goes about ordering his troops. The character in the film *Arsenic and Old Lace* who thought he was Teddy Roosevelt and spent half the film charging up and down the staircase, going off to battle was delusional. Meanwhile he was digging graves in the basement to bury his "troops" was how the two main characters of the movie used his delusion for their own purposes. While a delusion may look on the surface as if someone is just pulling your leg, it is a very fixed belief, and attempting to argue them out of their belief only results in anger, and often aggression, on the part of the client.

To better understand what it is like for the schizophrenic, imagine this scenario. You are hired at a school to be a teacher. You arrive the first day of school, go to your classroom, and start teaching. An hour later a person arrives and asks you what you are doing. You reply that you are teaching. They look at you with a concerned look, and ask your name. You tell them that you are John Smith, and they call you out of the class. In the privacy of the hallway, they tell you that you aren't the teacher, you are the janitor. You argue, no, you've graduated from Fuller University, you have a Master's Degree in Education, and you are the teacher. You offer to show your credentials, and they shake their head calmly, saying you need to come with them to Human Resources. You resist, demanding to know who *they* are. They firmly grasp you by the elbow, saying that is not important, and if you don't come with them immediately, they will have to call security. How would you feel?

Now let's take this one step further. Let's say you are allowed to teach the class, but meanwhile, someone has stuck an Ipod in your pocket and earbuds in your ears. However, each ear is hearing a different radio station while you are trying to teach. That is what it is like for a student in a class with auditory hallucinations. They have to concentrate on multiple voices—those in the real world, and those in their head.

Here is an example of just one of my lowest-functioning psychotic clients. Deonte came onto my caseload when his therapist left (is this beginning to sound redundant yet?). At that time he was around twelve. Deonte had suffered a traumatic brain injury as a child, and had a cognitive impairment as a result. He also was schizophrenic with delusions. However, his parents did not want him on medication, because his older brother, who was had been in our school as well, had a drug abuse problem and they thought the medication would lead Deonte to abusing drugs.

Working with Deonte was similar to working with an Alzheimer's patient. He was a likeable young man, and I never had any concerns about my safety when I was with him, but he had an impaired short-term memory. Therefore, he would ask a question, and five minutes later ask the same question. Our sessions were, to say the least, exhausting and felt unproductive. Deonte was unable to retain much academically.

Deonte had delusions that he was someone famous (most delusions are of this type), and he insisted on being called by that name, would write that name on his class assignments, and would not respond when called by his real name. Deonte also became physically aggressive when his delusions were questioned, which unfortunately, the staff at the school often did—not recognizing that his delusions were simply that—delusions. They were under the impression that Deonte was being defiant, and they were working "by the book" to correct his behavior. Unfortunately, psychotic symptoms cannot be corrected by behavior modification, only by psychotropic medication.

As Deonte got older, his hormones kicked in, and with it, his psychotic behavior worsened. His delusions became more entrenched, he heard voices telling him to harm others and himself, and he was being physically restrained on a weekly, if not daily, basis. After a particularly bad week, where Deonte had AWOLed from the school campus and the police had to be called, Deonte had another meltdown. He claimed the voices were again telling him to hurt himself and others, and I called the police for a 72-hour hold. The school staff informed the parents that until Deonte was properly medicated, he could not return.

Two weeks later Deonte returned. He wasn't angry with me for hospitalizing him, and stated the voices had gone away. His delusions eventually lessoned, and after a year on the proper medication, Deonte was able to participate in activities that he previously had been unable to do because of his AWOL risk. Deonte was happier, the staff could breath easier, and the parents were happier. Would Deonte ever have a job? Yes, a menial one, if he lives in an assisted living facility where he is transported to and from the jobsite. Deonte is capable of learning and doing simple repetitive tasks. Will he work in a movie theater running a movie projector like he wants to? No. Will he live on his own? No. Deonte, at age sixteen, is still like a young child cognitively and emotionally. And without his medication, he would be a danger to himself and others. But with his medication to stabilize his schizophrenia, he can function in school and a work place.

Are my sessions with Deonte any more productive? Not really. His short term memory problems can't be fixed with a pill. But Deonte is very attached to me and says "Hi Jillian!" every time he sees me on the school campus, and sometimes in session he talks about his frustration at not being able to return to public school. He shares his feelings about not fitting in or being teased. We discuss coping skills, even though I know that next week he will have forgotten the conversation, but at least in that moment, I feel like we are doing therapy.

An important note on safety when dealing with psychotic students. As I mentioned, Deonte was often being restrained, but once he was properly medicated his physical aggression diminished. *It is important to never touch a psychotic person on the back without making eye contact with them first.* You need to ascertain that they are lucid (grounded in reality) before touching them; otherwise, they may not be able to distinguish between you and their hallucinations and may harm you, thinking you are associated with the voices in their head. I had a client who was new to the school, and her IEP stated she was psychotic, had auditory hallucinations, and her coping skill was a time out. The teacher called me the day after I met Bridgit for the first time. The teacher was upset, saying, "Jillian, Bridgit says she's hearing voices!"

I was non-perturbed. "Yes, what do you want me to do about it?"
"Is she lying to me?"

Now I was exasperated because the teacher had failed to read an important IEP (as I mentioned, I always skim them looking for key words such as "psychotic"). "No. She's psychotic and hallucinates. It's in her IEP. Give her a time out, that's her coping skill. I'm not going to reward her symptoms by giving her a second session."

I made a mental note to check on Bridgit, and two hours later I saw her sitting with her peers at lunch, and I approached her, calling her by name. She turned, said hi to me, waving, and seemed lucid. I asked, "You okay?" and touched her on the shoulder in a reassuring manner. She nodded, and I went on my way.

Another important note when dealing with kids with psychotic issues—do not be afraid to ask them questions. It's similar to talking about sex. If you are frank and open, and don't make a big deal about it, the conversation won't be embarrassing for the children. If they sense you are uncomfortable, they will be uncomfortable. When I am assigned a student whose IEP states they hear voices, I'll ask straight out, "So, I see in your IEP that you hear voices. Can you tell me about that?" I need to ascertain how often they hear the voices, whether the voices tell them to harm themselves, or others, or both. I need to know what medication they are on, and what they do to cope with the voices. It's like suicidal ideation—you can't ignore it and hope that it will go away. It's a big elephant in the room, and it has to be addressed.

Another psychotic client of mine, who has been on my caseload since I've been at the school, is Ronny. Ronny lives in a group home, and it is honestly hard to gauge his intelligence. Sometimes Ronny can appear quite intelligent, and write things that are very eloquent and age-appropriate. At other times, his writings, like his speech, make no sense. While Deonte has a pretty good sense of self-esteem, and has strengths such as sports and can do certain things very well, Ronny has almost no self-esteem, and is almost always seen on the campus sitting by himself with a depressed look on his face.

It took several months before I realized Ronny was psychotic. Sometimes psychotic features appear immediately, in other cases they are more subtle. Ronny is in the latter category. Ronny appears to have a speech problem, but my guess is that the voices in his head are always competing for his attention and he hasn't learned how to tune them out when he is talking. The same may be true of his writing. As I mentioned, sometimes his writings are brilliant. Other times they are nonsensical.

Ronny will say things like, "Have you ever wondered who would win in a fight? A banana sandwich or an Oreo cookie?" When you ask him what he means, he'll just laugh, and start explaining it. Ronny has an extraordinary sense of curiosity, and is always asking questions about science—unfortunately, his curiosity does not stay grounded in reality.

Ronny also has delusions, but his take the form not of an identity, but of special powers. Ronny used to believe he could control the weather by nodding his head. So he was continually nodding his head when he walked around campus. Naturally, people thought he was odd for doing this. No one thought to ask him why he was nodding his head. One day I did and that was when I found out about his delusions. Now he does not nod his head, but holds his arms in a peculiar fashion, which makes him look like he's about to break out in a dance or is engaged in some sort of frozen statue game. Unfortunately, Ronny's delusions, just like Deonte's, makes him a target for teasing. However, Ronny has the added problem of not living with his biological family for emotional support, and therefore suffers from depression. Unfortunately, many kids in the foster care system do not get the right medication because the psychiatrists only spend ten minutes or less with the children while dispensing medication.

As a result, over one summer break, Ronny had a bad psychotic episode where the voices told him to hurt others and himself and he had to be hospitalized. He finally was put on the right medication. For over three years I had been telling his social worker that Ronny was psychotic and asking if he was on the appropriate medication, and I kept getting waylaid, being told she would "follow-up." Sadly enough, while Ronny is academically more capable of succeeding than Deonte, his prognosis is actually worse because of being in the system.

So what do Ronny and I do in his sessions? Primarily, we deal with his depression and feelings of isolation. We discuss ways for him to fit in. I can't deal with his delusions (the medication needs to address that), although whenever I pick him up for a session and his arms are contorted I tell him, "arms down" because we do work on him "fitting in" during session. When he is with me, he does not use his "magical thinking," but maybe that is because he feels safe. It is out there on campus where he feels threatened that his bizarre mannerisms manifest themselves the most—especially when on the yard.

Ronny also talks about his feelings about not having his biological family. Especially when the holidays approach. Mainly, I just work on his feelings of abandonment—given his level of depression and lack of support, all I can hope for is to give him lots of coping skills.

I do not want to give the impression that all schizophrenics are doomed to a life of institutionalization. These are two extreme cases, but the school I work at gets extreme cases. While in private practice, I did couples counseling with a schizophrenic and her husband, and they were there to try to find ways to cope with her schizophrenia. Where they lived there were no support groups. She was able to function at a low-paying job, they had children (she was worried about them getting schizophrenia from her since it's genetic), and their marriage was difficult, but that was why they were in therapy. But they had the support of their families. As I said, the school I work at tends toward multiple-diagnoses. If Deonte did not have a brain trauma injury, or Ronny lived with his family, both their outcomes would be more positive. But that is why this population is considered hard-to-reach.

A word of advice—many students who are psychotic, mentally challenged, or in other way low-functioning, come to school with unrealistic career goals. I have had students who were unable to read state that they planned to be a veterinarian. Other kids who can't do basic math plan to get jobs in retail working as cashiers. Year after year I will sit in the IEP and hear the teachers discuss the student's career objective, which is often completely unobtainable. Again, it is like the elephant in the room that no one wants to bring up. However, for the parents, to find out after their child graduates that their child is unable to get a job because they are unable to even fill out a job application, let alone fulfill their dream of becoming a doctor, it is frustrating and angering to have been kept in the dark for years about their child's limitations. There are ways to constructively work with a student and their parents to find ways of utilizing the child's strengths. If a menial job is all they can do, it is better to prepare them while they are still in school, rather than have them find out after they have left the environment where all the resources were available to them for free.

19

Mood Disorders

Depression and bi-polar disorders are common at my school. They are common among group homes and residential facilities. Usually, kids with these problems have another problem—such as a learning disability, were sexually abused, have ADHD, or as was the case with Ronny, my schizophrenic young man who hears voices and is delusional, may have significant other impairments to cope with. With Ronny, his depression came on gradually.

When he was first assigned to me, he was not depressed. He smiled all the time, and was always happy to see me (he still is, you just wouldn't know it to look at his body language). He had a friend that he was close to, and there was another higher-functioning student in the school who "looked after him." But then the higher-functioning student left for public school, and a semester later Ronny's friend got transferred to a different class. Suddenly Ronny felt alone; as though the world was against him. Also, his hormones kicked in and he had a growth spurt, and—well, to put it kindly, it did not treat him well.

So now when I see Ronny, he is always frowning, body slumped forward, he looks anorexic due to his lack of weight gain, and he is unable to focus on simple games that before he used to win. He is easily distracted. In other words, Ronny is suffering from a Major Depressive Disorder.

The following is a summary of the DSM-IV criteria for a Major Depressive Episode. There are other elements, but for the purpose of this discussion, the following is adequate. The DSM-IV also has sub-categories and codings for this diagnosis—for example, Major Depressive Disorder, Single Episode, Moderate—but again, in attempt to keep this simple, the following information will suffice. Again, as with schizophrenia, refer to the DSM-IV manual to see the complete details and coding specifiers.

DSM-IV Criteria for
Major Depressive Episode

Five (or more) of the following symptoms have been present during the same 2-week period and represent a change from previous functioning; at least one of the symptoms is either (1) depressed mood or (2) loss of interest or pleasure

a. Depressed mood most of the day, nearly every day (in children and adolescents can be irritable mood)

b. Markedly diminished interest or pleasure in all, or almost all, activities most of the day, nearly every day

c. Significant weight loss when not dieting or weight gain (in children consider failure to make expected weight gains)

d. Insomnia or hypersomnia nearly every day

e. Psychomotor agitation or retardation nearly every day

f. Fatigue or loss of energy nearly every day

g. Feelings or worthlessness or excessive or inappropriate guilt nearly every day

h. Diminished ability to think or concentrate, or indecisiveness, nearly every day

i. Recurrent thoughts of death, recurrent suicidal ideation without a specific plan, or a suicide attempt or a specific plan for committing suicide

There are two types of depression—situational and biological. Situational depression occurs after an event takes place in someone's life, and the person feels depressed for longer than most people would given the event, or feel more severely depressed than most people would. For example, the loss of a job, having to move to another state and leave all your family and friends behind, or the death of a family member (after the normal grieving period is over). In these cases, the depression is related to a specific situation, and usually resolves once the situation resolves. When the person who loses their job finds employment, gets financially stable again, and makes new friends in their work environment, their depression will probably be gone, and often without the use of therapy or medication.

Biological depression can be hereditary. If a parent has a tendency towards depression, then the client may have a tendency towards depression. While there may not be full-blown major depressive episodes, there may be a chronic mild depression (dysthymia) off and on throughout the person's life. People who have a genetic predisposition towards depression usually have a chemical imbalance in their brain and may require a very low dose of medication to stabilize them. Women who are genetically predisposed to depression are also more likely to suffer from post-partum depression.

When working with adolescents, one of the following questions I repeatedly get asked by teachers and staff is—how can you tell depression from normal adolescent "moodiness." When you work with a student five days a week, or live with them in a group home or residential setting—you'll recognize depression when you see it.

One more obvious example was Olga, around seventeen, who had a difficult home life, but at school she was very popular, and had a boyfriend outside of school. She was always seen at school during outside activities at a table surrounded by her friends, smiling and laughing. As things at her home grew worse, she started to withdraw, and kept to herself on the school grounds. She sat isolated from her friends, and was gaining weight. The kids started rumors that she was pregnant, but she was starting to eat out of depression. She already had a learning disability, and now was having problems concentrating in class due to her problems at home, and was falling behind, which made her self-esteem drop. After awhile, she had insomnia, weight gain, loss of interest in pleasurable activities, fatigue, inappropriate guilt, diminished ability to concentrate, and suicidal ideation. One day, my phone rang, and it was her teacher, saying that Olga wanted to see me. I had already seen her that week, but I knew that if she was requesting me, it had to be urgent. The teacher told me over the phone that my student had been crying for several hours. I said I was with a client, but would be there in approximately twenty minutes.

When I got to Olga's class, she was outside sitting in the hallway, crying. I sat next to her, and she said, "I think I need to go into the hospital again." We talked about it, and it turned out that Olga was afraid she would harm herself over the upcoming winter break. Olga had been receiving medication previously from a psychiatrist, and was nearly eighteen. We discussed her getting back on her medication instead, since she did not appear to have a plan or serious intent to harm herself. Since she was almost eighteen and was a danger to herself, she technically could consent to her own treatment without her mother's permission. As it was, her mother's signature for services was still valid, and Olga was seen on an emergency basis that day and started up with anti-depressants. I had Olga make a suicide contract with me, and we discussed coping skills for the holidays. Prior to the break I saw her twice a week for thirty minutes rather than once a week for an hour, just so that she would have that extra contact with me until the break.

When you see a student who is normally outgoing, social, and smiling, turn into someone who is isolative, guarded, and distracted—it could be many things, but one of the things that is necessary to investigate is the possibility of depression because there is always the possibility of suicidality—especially with adolescents who feel their lives are hopeless.

The one thing I've learned is to never ask, "Are you depressed?" If a student tells me they are depressed, that is fine. But many minorities, or lower-socio economic groups, associate depression with staying in bed all day and wanting to commit suicide. Males tend to associate the word "depression" with weakness. So instead I ask the following:

- ✓ How is your sleep?
- ✓ Are you skipping meals?
- ✓ Comfort snacking?
- ✓ Are you tired?
- ✓ Do you still play basketball? (Or whatever their hobby was)
- ✓ Are you still talking with your friends on the phone?
- ✓ Is it harder to concentrate in class than before?
- ✓ Does your mind wander?
- ✓ Do you feel like things are hopeless and may never get better?
- ✓ Do you feel irritable or angry most of the time?
- ✓ Do you cry more than usual?

Depression can also affect young children. The youngest example was a child who came onto my caseload at age six. When I asked him why he came to our school he explained he tried to kill himself at his old school by trying to jump out of a window. I thought he was exaggerating, and after our initial session, I looked it up in his IEP. He was telling the truth.

I had a young boy with Asperger's, named Joel, who came to our school and did not fit in. Due to his poor peer skills, he was a target for bullying and teasing. Unlike most children with Asperger's, he was not academically ahead of his classmates, due primarily from a refusal to do his school work. He had multiple other issues, including anxiety disorders, and hypochondriasis. He was whiny, and complained. He would refuse to do his class work, saying he had a stomach ache, so I would intervene stating, "If your stomach hurts, than you shouldn't draw during free time—you should put your head down and rest." His favorite activity was drawing—then he would pout and say his stomach did not hurt enough to keep him from drawing. So I would tell him if he was well enough to draw, he was well enough to do his reading. Everything with Joel was an argument. In some respects he was oppositional defiant, but I think it was more out of desperation to protect himself against teasing from his classmates.

He was at the school approximately one year, and I had never had a child on my caseload deteriorate in such a time. When he started at our school he made several friends in his first class, but that class was with severely disturbed, lower-functioning children who were not a threat to him academically, and did not bully him. Unfortunately, that was the wrong class for him. The right class academically for him, was the wrong class for him socially. The reality was that our school did not have the right class for his particular functioning level.

Joel got to the point where he was absent, his mother could not get him to come to school due to his fear of being teased and bullied, and he was crying in class daily. I asked him in therapy if he would be happier at another school and he said, "Yes." In his IEP I recommended a referral to another school. Within three months, Joel was sent to a more appropriate school (I'm sure my school did not appreciate me referring a student out but it was in the best interest in the child). During the IEP, Joel's mother told me how she was concerned about Joel, and had started him on anti-depressants, but she was worried about the side effects and horror stories she had heard.

I told her, quite honestly, that I was more concerned about his depression, and that as long as his medication was closely monitored by a psychiatrist, he should be fine. Given his level of depression, he was better off being on medication than not. In Joel's case, it turned out his depression was both situational (a reaction to the wrong school) and genetic. His family had a history of depression as well, which put him doubly at risk given his other psychiatric disorders.

On the more complicated side is the bipolar disorder. Why is it complicated? For a number of reasons. One, I have seen numerous children as young as six come into a school with a diagnosis of bipolar disorder, which I have not seen evidence of. The diagnosis of bipolar disorder in children has been a recent consideration, and quite frankly, I'm not certain I agree with it. I've never seen a child that fits the childhood criteria—usually their behavior mimics too many other disorders, such as oppositional defiant and ADHD. Sometimes these kids simply have really bubbly personalities—does that mean they need to be labeled as manic? Or is it that their parents do not know how to control them properly and went looking for medication to control them and the kid was slapped with a label to justify the medication?

Secondly, even among adults, bipolar disorders are commonly associated with other psychiatric disorders, including Anorexia Nervosa, Bulimia Nervosa, Attention-Deficit/Hyperactivity Disorder, Panic Disorder, Social Phobia, Substance-Related Disorders, and Borderline Personality Disorder. People with bi-polar disorders are more likely to have problems with school truancy, school failure, occupational failure, divorce, or episodic antisocial behavior.

Bipolar disorders, like the Depressive Disorders, have multiple types and codings (Bipolar I, Bipolar II, Cyclothymic, and the DSM-V will include Bipolar III). Again, I'm going to address the bi-polar disorder in general—but it should be kept in mind for clinical reference that a full coding could read Bipolar I Disorder, Most Recent Episode Depressed, Severe with Psychotic Features. For a complete listing of all mood disorders, please refer to the DSM-IV manual. As the DSM-IV does not acknowledge the inclusion of children for the bipolar disorder (only cyclothymic disorder), I will limit my discussion to adolescents which is covered by the DSM-IV.

DSM-IV Criteria for Manic Episode

1. A distinct period of abnormally and persistently elevated, expansive, or irritable mood, lasting at least one week (or any duration if hospitalization is necessary)

2. During the period of mood disturbance, three (or more) of the following symptoms have persisted (four if the mood is only irritable) and have been present to a significant degree:

    a. Inflated self-esteem or grandiosity

    b. Decreased need for sleep

    c. More talkative than usual or pressure to keep talking

    d. Flight of ideas or subjective experience that thoughts are racing

    e. Distractibility (i.e., attention too easily drawn to unimportant or irrelevant external stimuli)

    f. Increase in goal-directed activity or psychomotor agitation

    g. Excessive involvement in pleasurable activities that have a high potential for painful consequences (e.g., buying sprees, sexual indiscretions)

3. The disturbance is sufficiently severe to cause impairment in occupational functioning or in usual social activities, or to necessitate hospitalization to prevent harm to self or others, or there are psychotic features.

A bipolar disorder is diagnosed when a person has both a major depressive disorder and a manic episode over the course of time. Males usually have a manic episode first, while females usually have a major depressive episode first. People with bipolar disorders can cycle at different rates—some can by rapid cyclers, others can be seasonal cyclers, and some fall in the middle. Some can have mild depressive episodes and mild manic episodes. Others can have severe depressive episodes accompanied by suicidal ideation requiring hospitalization, and their manic episodes are associated with psychotic features such as delusions or hallucinations. It can pretty much be a mix and match.

Bipolar disorders can be one of the more difficult groups to work with for many reasons. One, it can be exhausting not knowing where your client is going to be emotionally week to week. Will they be depressed? Manic? Stable and able to function normally? The other problem is that many people with bipolar disorders tend to dislike taking their medications (which I will discuss at length in another chapter) and go on and off their mood stabilizers, causing limited progress in their treatment. Therefore, expect a lot of crisis management work associated with bipolar disorders.

Additionally, many people, especially adolescents, do not want to accept the fact that they have a "disorder" and will need medication for the rest of their life. The first bipolar adolescent I encountered in the school environment, Dana, was in the music business, and the first day I met with her she was suicidal. Fortunately for me, Dana was accepting of her disorder, and gave me her history willingly. She had multiple prior suicide attempts and hospitalizations. The PET team was called, and she did not require hospitalization. Dana was medication compliant and high-functioning. She was also a rapid cycler. One minute she would be writing a poem about death, crying, insisting she had to read it to you, and then two hours later she would be fine, all smiles and giggles.

Again, I get asked frequently, how is this different from normal adolescent drama? For one, with normal adolescent behavior, there is usually a trigger. When a girl gets weepy and tearful, it is usually over an external event—the break up with a boyfriend, a fight with a best friend, or something happened at home. With a bipolar disorder, like schizophrenia and depression, there is a chemical imbalance in the brain, and there does not need to be an external event to bring on the behavioral change. In addition, when a teenager breaks up with a boyfriend or girlfriend, this tends not to result in a reaction as extreme as suicidal ideation. The emotions of bipolars are out of proportion to the environment.

One bipolar on my caseload, who unfortunately lives in a group home, also has cognitive impairments. Because he is in a group home, the background information for him that I have is sketchy. All I know is that his biological parents have lost all rights to him, which suggests abuse and perhaps fetal alcohol or other drug-related issues at birth. Normally, bipolars have above-average intelligence. David, however, struggles academically and has short-term memory problems.

David is normally a somewhat manic, giggling, high-spirited boy. However, he has his down periods, and these can last for one to two weeks. During these periods, he will refuse to do any school work, his body language says "keep away," and he becomes sullen, uncommunicative, angry, and physically aggressive. However, in all the years I have worked with David, he never became so depressed as to be suicidal.

One day I went to pick him up, and he was reluctant to come with me. However, his teacher encouraged him to come, saying to me, "Maybe you can do something with him." When we reached my office, David went off on a tirade about how he didn't care about anything. He didn't care about what happened, he didn't care about school, he didn't care about me, he didn't care about whether he lived or died. He went on to say that maybe he'd be better off dead. I asked him if he thought about how he'd kill himself, and he immediately rambled off several ways he could kill himself once he got home from school. I immediately asked him, "Are you on your medication?"

When dealing with children or adolescents who have cognitive deficits, when asked a question they do not quite know how to answer, they will sometimes confabulate an answer. This is not the same as lying. It's a result of shame over not being able to clearly answer a question, so they answer it to the best of their abilities, filling in the details with what they think they should say, rather than admitting that they don't know the answers (you see similar examples with the elderly who have dementia or Alzheimer's).

David gave me a convoluted answer regarding his medication that made no sense, but the end result was that he had stopped taking his medication several weeks ago. Normally, I would make a suicide contract, but with a student in a group home situation (and I knew he did not have a good relationship with the staff at the home), and given his cognitive impairment, I called the police for a 72-hour hold. Given that he had multiple plans, multiple means, and had never in all our years together discussed suicide, I was not taking chances.

After David was taken to the hospital, I was finally able to reach the group home manager. Turns out, she had been busy, and had not been able to refill David's prescription. *She just did not think it was important.* I reiterated that without his medication, incidents like this would happen again, and told her she had to have the medication filled in order to get David out of the hospital and back at school.

On the other spectrum, is Kelly, who is academically bright, but refuses to acknowledge she has a bipolar disorder. Every three months, like a clock, Kelly has a major depressive episode where she is crying, despondent, wants to go back on medication, agrees that she has a problem, and "doesn't care about anything." Then once she's been on the medication long enough to be stabilized, she says, "I'm not bipolar, I'm just manipulative. I can control my behavior if I wanted to." She stops her medication, and in a few months time, is out of control, the police are at her house, she's skipping school, falling behind in class, in study hall and on LOP every week, and all the staff keeps clear of her. When she's manic, she gets her hair changed to a wacky color, gets more tattoos and body piercings, stays up all night, and sleeps in class because she's going to bed at 3:00AM.

Then Kelly hits her depressive cycle, and decides she has a problem and wants to go on medication again. I have spent the past two years with Kelly simply trying to get her to recognize her disorder. When she states things in session such as "I don't need medication," I write it in her file on a piece of paper with the date and time. Then the next time she wants medication, I pull out her file. Eventually Kelly was able to progress to the point where she could *admit* she might have bipolar disorder.

These kids have so many labels in their life. "Trouble-maker," "bad kid," "drug-addict," that they do not want one more label. So for teens like Kelly, I try to use real-life celebrities that they might be able to relate to so that some of the stigma is taken away. I point out the Linda Hamilton (*The Terminator*) and Carrie Fisher (*Star Wars*) have bipolar disorders. I try to accentuate the positive—that people with a bipolar disorder tend to be creative, intelligent, and that if properly medicated, can be very productive and successful. Carrie Fisher wrote a book about her disorder, *the best awful,* which I recommended to Kelly to read so that she could see how someone else struggled with the same things she was struggling with.

Kelly does have a lot of positive traits. She's intelligent, articulate, motivated, creative, driven, and goal-oriented. Unfortunately, when she's manic—all these positive traits disappear into a cloud of dust, and it's like "where's Waldo?" You can exhaust yourself looking for the picture to materialize, or you can wait for the cycle to run its course.

The most difficult challenge in working with a bipolar client is getting them to recognize what is a result of their disorder, and what is not. These kids have to cope with a chemical imbalance on top of dysfunctional homes, feelings of abandonment, gang environments, being on probation, living in foster care—sometimes denial is the easiest path. As one teacher said to me when trying to figure out how to cope with Kelly's constant mood swings, "Pavlovian methods won't work." No, but a nice dose of Depakote might.

20

The Kitchen Sink

So what about the psychiatric disorders that do not fall into the category of a pervasive developmental disorder, psychotic disorder, or mood disorder? Well, the bulk of what you will deal with among children (aside from specific learning disabilities) tend to be oppositional defiant disorder, conduct disorder, and attention-deficit/hyperactivity disorder (ADHD).

Many of the students at my school have ADHD. I've often joked that Autism is to the 21st century what ADHD was to the 1990s. Two decades ago, every kid in school with any behavioral problems received the ADHD diagnosis. Now Autism is being over-diagnosed. The problem is, now I see kids without an ADHD diagnosis, who clearly have the disorder. Unfortunately, that is the pattern—a new disorder gets over-diagnosed, then clinicians are reluctant to use the diagnosis and it gets overlooked. Below is the DSM-IV criteria for attention deficit/hyperactivity disorder. For children who only have the inattention part of the criteria, they would be diagnosed as ADD.

DSM-IV Criteria for
Attention—Deficit/Hyperactivity Disorder

Either six (or more) of the following symptoms of **inattention** have persisted for at least 6 months to a degree that is maladaptive and inconsistent with developmental level:

1. Often fails to give close attention to details or makes careless mistakes in schoolwork, work, or other activities
2. Often has difficulty sustaining attention in tasks or play activities
3. Often does not seem to listen when spoken to directly

4. Often does not follow through on instructions and fails to finish schoolwork, chores, or duties in the workplace
5. Often has difficulty organizing tasks and activities
6. Often avoids, dislikes or is reluctant to engage in tasks that require sustained mental effort
7. Often loses things necessary for tasks or activities
8. Is often easily distracted by extraneous stimuli
9. Is often forgetful in daily activities

Either six (or more) of the following symptoms of **hyperactivity-impulsivity** have persisted for at least 6 months to a degree that is maladaptive and inconsistent with developmental level:

1. Often fidgets with hands or feet or squirms in seat
2. Often leaves seat in classroom or in other situations in which remaining seated is expected
3. Often runs about or climbs excessively in situations in which it is inappropriate
4. Often has difficulty playing or engaging in leisure activities quietly
5. Is often "on the go" or often acts as if "driven by a motor"
6. Often talks excessively
7. Often blurts out answers before questions have been completed
8. Often has difficulty awaiting turn
9. Often interrupts or intrudes on others

Kids with ADHD are usually academically capable of being at grade level. Unfortunately, one of the reasons that ADHD usually gets overlooked is because it is associated with other psychiatric disorders such as Oppositional Defiant Disorder, Conduct Disorder, Mood Disorders, Anxiety Disorders, and Learning Disorders. One of the most complicated cases I ever worked with was a young man named Gabriel. I was the supervisor to his one-to-one aide, and would observe him and his aide two hours a week at his SED class. Gabriel was academically bright, but suffered from social phobia, obsessive-compulsive disorder, and when his anxiety was at its worse, he pulled people's hair. While watching Gabriel work in class the very first week, what I saw was his being easily distracted. He was off-task, interrupting his teacher, out of his seat (usually under it or on the table—not hiding, but more in a gymnastics hyperactive way), busy examining non-existent lint on his clothing, distracted by a spider, and generally looking everywhere in the class except on the piece of schoolwork in front of him.

That evening I called his mother to report what I had observed, and I tentatively asked if anyone had tested him for ADHD. It turned out he had been diagnosed with it, but was not on medication for it, because the psychiatrist was busy adjusting his medication to control his anxiety disorders first. After a discussion with the mother and the teacher, we all agreed that unless Gabriel was put back on the ADHD medication, all the redirection and token-reward system in the world would not help Gabriel focus in school. ADHD, like schizophrenia and bipolar disorder, is a result of the chemical imbalance in the brain. Older adults can learn coping skills to adapt to the ADHD with the use of medication, but to ask a seven year old to do this—especially when they are already dealing with multiple other psychiatric problems, is like asking someone with Alzheimer's to stop forgetting things. It is just not possible!

So, as someone working in a classroom with a child, how do you differentiate between a learning disorder and ADHD? The two are easily confused because they both can result in poor school performances. However, there are subtle, but distinct differences, if you know what to look for. Usually, kids with ADHD have a higher level of self-esteem than kids with learning disabilities. Children with ADHD know they can do the work, they are just not doing it. Not out of defiance, but because their brain is busy interfering with their ability to stay focused. But deep down inside, these kids know they are smart.

If you give a game to a child with ADHD to play (often referred to in school lingo as a "preferred activity") they will suddenly be able to concentrate and stay focused, especially if they are in a one-to-one situation with no distractions around them. It is in class (or when they are not on their medication), that they are easily distracted and start staring into space, fiddling with their pencils, shoes, erasers, doodling on their paper, or watching to see what their peers are doing. Next thing you know, their lesson time is up and they have not completed their work.

One example is Michael. Michael is only eight, but is smart enough to play games that I play with middle-school children. He learns easily, is boastful, works above his grade level, and is more or less well liked by his peers. Then some days he comes into my office and he is unable to stay focused while building a puzzle or doing an activity. He gets distracted by the information I keep on my bulletin board. "What's that?" He's distracted by someone out the window. "Who's that?" He starts looking for one piece of the puzzle, then gets distracted by another game or object. Then he starts singing a song. Then he has to play with his clothes. Michael can easily build a 100-piece puzzle (with a little help from me) in one session. When he is off his medication, all he can do is put together about 25 pieces. However, when Michael is on his medication, and he is in class, he is one of the first students to complete his work.

Then there is Keith. I have been working with Keith since he was nine, and he is now twelve. Keith has always been behind in his reading—it is his weakest subject. When Keith is in his office with me, he likes to draw and color. He is focused, and can do very elaborate artwork and does not fidget or get distracted while he works. Unfortunately, in his academics, Keith is falling further and further behind his peers. He physically acts out, and his self-esteem is very poor. He is larger than his peers and uses that to make up for his academic shortcomings ("I may be dumber than you but I can beat you up!").

In class, I often see Keith sitting slumped down in his seat, refusing to do his work. However, he is not staring at the other students, or fidgeting with objects. He is dejected, with an angry look on his face. Anger outward is a sign of depression. One day I came to get Keith, and he was working on math. The teacher said that if Keith did not complete his math, he would not earn his activity for the day. I told Keith to bring his math, and the aide in the class said to me, "Maybe you can help him. I tried, but he won't listen to me."

Keith was working on basic single-digit multiplication, and every one of his answers was wrong. Using my white-erase board, I showed him how to do the multiplication using sticks. He quickly figured it out, and did the entire work sheet during his session. Again, I do not usually make contracts with students, but in this case, because he spent his entire session doing school work, I told him that if he did the entire page, I would give him an eraser (the eraser on his pencil was gone).

Keith loves new pencils and erasers for his artwork—they are like gold to him. His eyes lit up, and he happily took the eraser back to class.

In working with Keith, I noticed he kept inverting the numbers. He would say 21, but write 12—or he would say 86, but write 68. Even when I would point this out to him, he would look at the number again and again, and insist it was right. I went back to the class and explained to the aide that part of Keith's problem was he had dyslexia. It turns out that the method I had used (counting sticks) was exactly what the aide had attempted to get Keith to do. What Keith needed was after school tutoring with a one-on-one to catch up academically. He was smart enough to grasp concepts, but because he was so academically behind, his self-esteem was in the toilet. In class, he felt too humiliated to even try, and often gave up as soon as the assignments were handed out. There was no signs of ADHD—just incredible low self-esteem resulting from a simple learning disorder, in this case dyslexia, that he could with time, learn coping skills to accommodate.

As I mentioned earlier, the majority of the boys at the group home fell into the oppositional defiant disorder category—as did several of the students whom I worked with at Watts. This is what I call the chronic toddler stage. One teacher I worked with referred to it as the "spoiled brat syndrome." The problem is that the behaviors are severe enough to disrupt the child's ability to obtain an education and often interferes with their ability to form friendships. These are the students who fall into the "mild" bully category. They will tease and annoy their peers, and while they *can* make friends, they have difficulty sustaining friendships. Deep down, they know their behavior is bad, and usually have some level of guilt about their behaviors.

DSM-IV Criteria for Oppositional Defiant Disorder

A pattern of negativistic, hostile, and defiant behavior lasting at least 6 months, during which four (or more) of the following are present:

1. Often loses temper
2. Often argues with adults
3. Often actively defies or refuses to comply with adults' requests or rules

4.   Often deliberately annoys people

5.   Often blames others for his or her mistakes or misbehavior

6.   Is often touchy or easily annoyed by others

7.   Is often angry and resentful

8.   Is often spiteful or vindictive

An example of this would be Damion, who I started working with when he was ten. He came to the school with a history of encopresis, and had horrible hygiene issues. Damion turned out to be academically on task, and had perfect school attendance, but in class he often defied the teachers, and would sit and stare into space. He instigated his peers for no reason, blamed everyone else for his behavior, and would argue with the staff whenever he was redirected.

While Damion's behavior at school was not terribly out of control, we learned from his parents that the reason for his poor hygiene was that Damion refused to do anything he was told at home. He refused to shower, brush his teeth, go to bed, do his chores, and would deliberately annoy his siblings. If he perceived a sibling to get preferential treatment, he would wait for the appropriate moment to trip or otherwise "accidentally" hurt them.

Damion was referred to a medical physician to rule out any medical reason for his encopresis but none was found. His behavior had started around age six, and was causing him further alienation with his peers due to his smell. The more discussions I had with him about his hygiene, the more he shut down emotionally. Yet, whenever Damion came with me to a session, he always held the door open for me, saying "Ladies first." He always said "please" and "thank you." His social skills were excellent. His academics (when he did his work) were excellent. He was completely capable of returning to public school except for his defiant behavior. I think, on some level, his shitting himself, to be perfectly blunt, was his ultimate level of defiance.

Oppositional defiant disorder is often a precursor to conduct defiant disorder if it appears at an early age. However, the two are extremely different, and once you have seen a conduct disorder—and yes they can exist in a child under the age of ten, you will never miss the diagnosis again. To put it simplistically—adolescents with a conduct disorder generally grow up to be adults with an anti-social personality disorder. As with the oppositional defiant disorder, there is significant impairment in functioning in school, social relations, and occupational settings. However, there are other significant differences. These are the "major" bullies. They use physical aggression to intimidate peers, usually have no friends, show no remorse for their actions, and may boast about their behaviors. First, let's look at the criteria.

DSM-IV Criteria for Conduct Disorder

A repetitive and persistent pattern of behavior in which the basic rights of others or major age-appropriate societal norms or rules are violated, as manifested by the presence of three (or more) of the following criteria in the past 12 months, with at least one criterion present in the past 6 months.

1. Often bullies, threatens, or intimidates others
2. Often initiates physical fights
3. Has used a weapon that can cause serious physical harm to others
4. Has been physically cruel to people
5. Has been physically cruel to animals
6. Has stolen while confronting a victim
7. Has forced someone into sexual activity
8. Has deliberately engaged in fire setting with the intention of causing serious damage
9. Has deliberately destroyed others' property (other than by fire setting)
10. Has broken into someone else's house, building, or car
11. Often lies to obtain goods or favors or to avoid obligations
12. Has stolen items of nontrivial value without confronting a victim
13. Often stays out at night despite parental prohibitions, beginning before age 13 years
14. Has run away from home overnight at least twice while living in parental or parental surrogate home (or once without returning for a lengthy period)
15. Often truant from school, beginning before age 13 years

How then, does the above criteria differ from the life style of a gang-banger? Good question. First of all, remembering what was said above, gang members do have friends and are capable of forming lasting relationships. They are fiercely loyal and protective to their families, siblings, and often join gangs because they want to emulate a family member. When a gang-member has to commit a crime (like my student Enrique who was often asked to use a gun to do something, and would refuse and get into a fight), this is often part of the "initiation" into the gang, and while the child may comply, they usually feel remorse over their actions. There is a difference between a group acting as one—in a gang—committing a crime such as robbery or grand theft auto, and a single person operating alone. Gangs are operating on a territorial, retaliation pattern of behavior. A child with a conduct disorder is operating alone, and therefore has no theoretical reason for their behavior. The gang-members I have worked with have all shown remorse and guilt over the activities that they have committed. When telling me about their criminal acts, they have been ashamed, worried about what I would think of them or that I would judge them. When a person with a conduct disorder tells you they have raped someone—they are going to be boastful and puffed up. It is as though the part of their brain that knows right from wrong is missing. So while it is easy to look at the above criteria and think gang-members have conduct disorders, the DSM-IV needs to modify the criteria to clarify that the activity is engaged in by one person alone as opposed to a group, to differentiate from gang-activity.

The first conduct disorder I ever worked with was a young girl, Jasmine, who was not at our school long before being placed in a level 14 (locked) facility. Jasmine was continually initiating physical fights with her peers and staff. Her retaliation to events was often distorted out of proportion to the event. If someone reprimanded her, she would throw a desk over. On one occasion, she got into a hair-pulling match with a student, and two staff had to break up the fight. One of the staff had to be hospitalized, and the teacher had bite marks on both of his arms that looked like he had been assaulted by a wild animal. When I questioned Jasmine about it, she said with a smile, "They deserved it." When I pointed out that she started the fight, she made up an elaborate lie to avoid taking responsibility for her actions. When I pointed out the staff had to be hospitalized due to her actions, she only shrugged, showing no remorse.

I have only worked with three conduct disorders—and I always make certain that a therapist that I trust is in the office next to me before I pick them up so that I can shout for help if necessary while in session with the child. Fortunately, these children rarely last for more than a few months in a facility less that a level 14 before being transferred to a locked facility.

In the group home, there was a man with a conduct disorder (again he was there briefly before being hauled away by the police) who spent all of his time discussing how he was going to kill people. He was busy trying to figure out how to make bombs, he wanted someone to break into the group home so he could kill the burglar with a baseball bat, he wanted a sexual predator to assault him in the park so he could kill him. All he talked about was death and mayhem. The entire time he would discuss this, he would smile, and laugh manically. I told him he used all this death and destruction talk to keep people at an emotional distance, and to stop with the bullshit talk and tell me what he was really feeling. Whenever we got close to discussing something that scared him, the talk about hurting others would start up again. Needless to say, feelings were outside of his comfort zone.

It was like watching a young Hannibal Lector. Interestingly, in our first session he gave me this long, tangential story about his childhood, which was primarily lies. He seemed to be testing me to see if I'd feel sorry for him. I told him to "cut the crap and tell me something real, from his heart." He said I was "mean and sarcastic."

"Oh, I'm sorry, were you looking for someone who was going to say boo, hoo, have a tissue. You poor little victim?" The boy had stolen into his parents home with a gun and hocked their possessions for drugs. "I'm not a touchy, feely therapist, so if that's what you were hoping for, you're going to be disappointed."

At that, he stared at me, and I earned his respect—or at least as much respect as he was capable of. That didn't mean we got along well, or that I looked forward to our sessions, but I had set incredibly high boundaries, and made it clear he was not going to get through them. He might have manipulated his previous therapists, but not me. He did not know how to deal with that, and I let him sit in his discomfort and struggle with it.

Kids with conduct disorders have immense self-esteem issues. They feel superior by conning people, and getting away with crimes.

That's all they've got. If they can outwit the police and the system, then they must be smarter than all those psychologists and doctors in the mental hospitals and juvenile halls that they have to deal with. So when working with a kid with a conduct disorder, it is imperative to keep control. If you have nothing self-assured to say—simply look at them calmly and say nothing. While this will cause them to lose control, that is how you can gain control. Half the time they'll start twitching, screaming, "Why are you staring at me! Stop staring. Why don't you say something?" To which I reply with a nonchalant shrug, "What do you want me to say?" Or, I'll flat out turn the tables and say, "Why should I say anything? You keep saying you know more than me. Obviously I can't help you, so let's go back to class. We're done here. You already know everything." At which point they will get upset because they do not want their counseling time to be over.

With low self-esteem kids you normally want to *empower* them as much as possible. With conduct disorder kids—you want to *disempower* them by breaking them down and getting them to admit one thing and one thing only. *That they have a problem.* Conduct disorders make alcoholics look like functioning corporate executives. Unfortunately, because people with a conduct disorder generally grow into adults with a personality disorder, it is unlikely that they will change unless you catch them early—around age eight. And as sad as this may be, I have seen them start down that path that early.

21

The Obsessive Picture

As mentioned in the chapter on Pervasive Developmental Disorders, children typically go through obsessive phases. This has been a phenomenon that toy companies, the music industry, the entertainment industry, and most currently, the video industry has taken advantage of for decades. Starting with the Beatles through to 'N Sync to Hannah Montana, marketing companies have profited from the fact that children will obsess about a celebrity and make as much profit off of this as they can in as short a time as they can. Fads such as Pogo Sticks, Cabbage Patch Dolls, skateboards, X-Box, Pokemon, and Star Wars have come and gone, and sometimes come around again.

The one psychiatric diagnosis which I have not yet mentioned is Obsessive Compulsive Disorder. There is a simple reason for this—in all my years of working with children in restricted settings and private practice, I have never encountered a child with OCD. According to the DSM-IV, OCD typically begins in adolescence or early adulthood. In their work, *Obsessive-Compulsive Disorder Casebook,* Drs. John Greist and James Jefferson present sixty cases, of which 20 percent are ages eight to sixteen. I have worked with over 500 children in various school and private practice settings and never come across any cases of OCD; but I contend that that is not my specialty. Additionally, I suspect that OCD is not as prevalent in the lower socio-economic groups served by the schools I have worked with. OCD tends to be associated with Major Depressive Disorder and Tourette's Syndrome. However, I have had adult clients with OCD in private practice, and am very aware of the difference between a normal childhood obsession—or an atypical obsession associated with another psychiatric disorder, and a true Obsessive-Compulsive Disorder. How, then, to tell the difference?

Obsessive-Compulsive Disorder is an anxiety disorder. Most people have in their mind that OCD consists of people who are phobic of germs and perpetually wash their hands. While this can be one manifestation, it is over simplified, and the disorder can take many forms. The film *As Good as it Gets* exemplified OCD brilliantly with Jack Nicholson's character whose life was so overruled by his compulsions he was unable to form meaningful relationships with others. People with OCD tend to be unhappy with their rituals and compulsions. They are aware that their behaviors are not normal.

While their compulsive behaviors or obsessive thoughts may have provided relief from their initial anxiety in the beginning, their obsessions and compulsions eventually take over their life, and provide no relief from the anxiety. While the following analogy may sound simplistic, it is somewhat accurate. OCD is similar to a drug addiction in that the ritualistic act (or thought) may provide relief the first time it is performed, but the more it is performed, the less relief it provides, causing the client to have to perform additional rituals. Just as a drug addict initially gets high from one pill, eventually they have to take two, three, four pills—and eventually the pills do not work at all.

OCD, like schizophrenia or bi-polar disorder, is a neurological based problem. PET scans of the brain have shown that the orbital cortex of the brain has higher activity in OCD patients than a "normal" brain. As a result, OCD is usually treated with a combination of psychotropic medication and psychotherapy.

While this book is on working with children who often have dual diagnoses, because they do often exhibit obsessive behaviors—especially those associated with Autism and Asperger's, I want to touch on OCD for comparison purposes. Again, refer to the DSM-IV for the complete criteria.

DSM-IV Diagnostic Criteria for Obsessive-Compulsive Disorder

Either obsessions or compulsions (obsessions are defined as 1 thru 4):

1. Recurrent and persistent thoughts, impulses or images that are experienced, as some time during the disturbance, as intrusive and inappropriate and that cause marked anxiety or distress.

2. The thoughts, impulses or images are not simply excessive worries about real-life problems

3.   The person attempts to ignore or suppress such thoughts, impulses or images, or to neutralize them with some other thought or action

4.   The person recognizes that the obsessive thoughts, impulses, or images are a product of his or her own mind.

Compulsions are defined by 1 and 2:

1.   Repetitive behaviors (e.g., hand washing, ordering, checking) or mental acts (e.g., praying, counting, repeating words silently) that the person feels driven to perform in response to an obsession or according to rules that must be applied rigidly.

2.   The behaviors or mental acts are aimed at preventing or reducing distress or preventing some dreaded event or situation; however, these behaviors or mental acts either are not connected in a realistic way with what they are designed to neutralize or prevent or are clearly excessive.

In order to best illustrate what does constitute OCD, I will talk about an adult client I saw briefly in private practice. Susan was a "checker," meaning that she had to check things before leaving her home. She had to check that the stove was off, that the iron was not on, that appliances were unplugged, and that the door was really locked. Unfortunately, after locking the door for work, she had to go back inside and check everything again. Susan had lost several jobs because she was always late for work and was currently unemployed. Susan obviously was unable to comply with therapy, because she was unable to make her appointment times. Susan illustrates a more conventional OCD patient.

A less conventional OCD patient was Irina. Irina worked for a successful advertising agency, and to all outward appearances she was very well put together. However, Irina's life was regimented by lists. She made a list of all the meals she would eat for two weeks in advance, what she would wear everyday, what she would do. On her weekends, she made a list of her errands, down to the minutes—how much time it would take her to get to each errand, the time spent at the errand, or so on. If Irina was unable to eat at the same restaurant, eat the same meal, in the same seat, she would have a panic attack. Irina was often invited to lunch by her co-workers and to after-work events, and declined, because these things were not on her list, and she had to stick to her schedule. Irina would have loved to accept these invitations, but was unable to break out of her rigid system. Unexpected events caused her distress to the point where she would cut herself to relieve her anxiety.

In addition, Irina was a hoarder to a mild degree. She stockpiled items in bulk in case of emergencies. She had toothpaste, shampoo, and other essentials to withstand the end of the world. The problem? It was putting her into financial debt. In addition, she was buying things she did not need. If Irina had coupons for items, she felt compelled to buy the product, even if she already had twenty of the items at home. When asked why, she replied, "If I don't use the coupon it's throwing away money." Irina was unable to see that buying something she didn't need to save twenty cents was actually costing her—she *had* to use the coupon. She stockpiled foods in case of an earthquake; however, they were foods she did not like and would never eat.

No one in Irina's family knew of her problems. She had no friends to tell her pain to. Her boyfriend knew only the slight superficial extent of her problems, and she feared losing him. Her therapy sessions with me reminded me of Catholic confessionals where every week she would confess another secret she had been holding in for her adult life, waiting to see how I would react. Waiting to see if I would give her a verdict of "crazy." The more I accepted her secrets without judgment, the more secrets came tumbling out as though a Pandora's box had been opened.

After working with Irina for a year, getting her to become medication compliant (which she was not when she started treatment with me), and encouraging her ever so slowly to step out of her comfort zone, Irina started to socialize, eat different foods, and eat at different restaurants. She eventually terminated therapy because she took a job that she wanted in a preferred field which conflicted with her appointment time. On a follow up call a year later, Irina was doing well, taking risks in her life, and was happier.

So what about my students? As I said, obsessive behaviors is common to childhood. The difference is their obsessions or compulsions usually do not interfere with their ability to form relationships or attend school. With time, their obsessions move on to other obsessions. When a child is unable to play a video game, he may tantrum, or pout, but it is because he usually does not want to do his homework—not because being deprived of the video game is causing him obsessive thoughts or anxiety. Children often do not want to give up their toys, and have a cluttered bedroom, but if they refuse to give up their toys to the point where it is causing them anxiety, this may be a sign of early hoarding, which is often how OCD manifests itself in children. Usually, a teacher or therapist can tell when a behavior is normal, or anxiety based, by simply asking the child to perform a task differently. If the child is unable to perform a task outside a ritualized manner, then OCD is probably at play.

Children, especially pre-adolescents, are driven by self-image. They want to fit in. If everyone is watching one particular television show, they will watch it too—not because they like it, but because they want to have something to share and talk about when everyone else talks about it. They will listen to the same music, wear the same clothes, and play the same video games. That is why the stereotypical "nerds" are just that. They are the ones who generally do not follow the herd mentality, and are consequenced by being labeled an outcast or misfit by their peers.

One high-functioning Asperger's child I worked with became obsessed with playing scrabble with me, because he kept losing. He played over thirty games in a row (counting each week) until he finally won. Then he moved on to chess. Why? His teacher played chess. Often, these students will attach to their adult role models and want to mimic their behaviors. It is as if they are adrift and are searching out for their own identity, and because they are unsure of who they are—and often who they are has not been good enough—they cling to the identities of those around them whom they admire. I was good with words, so my client became fixated with scrabble. His new teacher was good with chess, so he became fixated with chess. As my client grew more comfortable in his own skin, he developed his own interests and identity, and was able to let go of his obsessive behaviors. Not completely—but noticeably.

22

Concerta, Abilify, and Risperdal, Oh My!

There is a tendency, among the population of hard-to-reach youth, to be a general lack of medication compliancy. There are several reasons for this. With certain demographics, there is a misunderstanding about the use of psychotropic medications, and many parents think that putting their children on medication leads to drug abuse. Additionally, as many of the cases I have presented have illustrated, foster parents simply do not want to take the time to deal with additional trips to the doctor to obtain the medication, or to follow-up with the physician if the medication is not working. Then there are the children themselves.

Just as these kids do not like all the labels they receive—"special education," "learning disabled," "bipolar"—or sometimes their parents just plain tell them they are "bad,"—these kids see the pills as further proof that they are "crazy," and so they refuse to take their meds. One of my ADHD clients had very poor medication compliancy (as did his parents who felt the medication was too expensive). I could always tell when Charles was off his medication. He would always insist that he was, and I would reach over for my Catch the Match game, and say, "Okay, let's see how well you do on this." At that point, he would stand up, screaming, "All right, all right! I didn't take my meds!" He knew that if he was on his medication he could focus on the game and win, but he also knew that if he wasn't on his medication he would not be able to focus—and he knew that I knew.

When I have children who have issues with taking their medication there are several ways I explain the importance of the medications to them without getting technical. I'll ask my client, "Does anyone in your family have diabetes?" Another good example if they answer no is to ask about high blood pressure. However, because these children come from families with poor nutritional habits, they almost always answer yes to the diabetes question. I'll ask follow up questions, such as who is the relative? How close are they to that relative? Then the conversation will go something like this:

"So your grandmother has to take medication for her diabetes, right?"

"Yeah. She takes something; I don't remember. And she gets her blood checked with a little machine."

"Yes, that's right. She takes medicine because her body doesn't make enough insulin, and her body's blood chemistry is out of balance." I pause while they digest this. "Do you think your grandma's crazy because she has to take medication?"

They shake their head. Maybe they'll ask a question. "Is it true people with diabetes can die?"

"Well, anyone can die. But yes, if your grandmother doesn't take her medicine, she can have problems with her eyesight, her heart, and diabetics often have problems with their legs and feet." Again I wait for them to digest this. If they ask questions I answer them. "Now, you'd be upset if your grandmother stopped taking her medication, wouldn't you."

"Yes!"

"Okay. Now, don't you think your grandmother would be upset if she knew that *you* weren't taking your medication? Your ADHD is just like your grandmother's diabetes. Your brain isn't producing enough of a certain chemical, and that's why you need to take your medication. It doesn't mean that you're crazy. You're just like your grandmother. Your body is out of balance, and this medication corrects that balance."

Usually after I have this conversation, depending on the age of the child, and what their problem is (ADHD is different from schizophrenia or bipolar disorder), there will be a long conversation on how long they will need to be on the medication, and how they don't like the side effects. However, the end result is that the child usually feels less "crazy," once it is explained in a way that normalizes it for them, and you can associate their problem with a family member that they are emotionally attached to.

So how do psychotropic medications work? Medications have been around since around the 1950s, and are continuously being refined. The medications available now compared to the early drugs (which predominantly sedated people to the point of eliminating *all behaviors*), now target specific parts of the brain and address neurotransmitters and specific chemicals. Early anti-depressants consisted of MAO-inhibitors which inhibited the action of monoamine-oxidase. Tricyclic anti-depressants came around the 1980s, and then came the SSRIs (selective serotonin reuptake inhibitors around the 1990s (these drugs can also target dopamine and norepinephrine).

Through the use of PET scans, researchers have been able to study the brains of people with various psychiatric disorders (bipolar, schizophrenia, obsessive-compulsive disorder) and determine which parts of their brain differ from the general population. Usually researchers find that one area of the brain is either under-active or over-active. It is now actually possible, with depression, to have tests done telling you exactly what your serotonin levels are (if you want to pay for it). While there is some controversy on how effective these tests are, my guess is that in the near future they will be refined to the point that the tests can pinpoint which chemical (serotonin, dopamine, or norephinephrine) is too low in your body.

Why would anyone want to do this? Well, because as you will see, the brain is filled with so many chemicals, neurotransmitters, and nerve cells, there are literally dozens of types of medication available for every disorder. This is because each medication targets a different chemical in the brain. Unfortunately, because there are so many possibilities causing a biological depression (as opposed to situational depression), a client may have to try five or more anti-depressants before finding the one that works for them. This is why people often give up on medication before having any results.

Additionally, as medications become available to the general public, the prescribing doctors are finding that certain medications address multiple symptoms. So while Paxil, for example, was originally developed as an anti-depressant, they found that it also addressed anxiety (this is why you now see commercials for Paxil for "social anxiety.") Cymbalta is marketed as treating depression and the reduction of pain associated with certain medical conditions.

There is a lot of news about the use of psychotropic medications associated with suicides and children. However, the news tends to glorify the sensational. If a plane crashes—it makes the news. If in the same day 1,437 planes landed safely, that does not make the news.

All medications have side effects, and with normal adolescents (those in public school, an intact family, and a good social support system), it is wise to watch out for suicidal behaviors if your child is depressed enough to warrant psychiatric medication. In a population where the children are in an unstable home, a special education setting, may not have good verbal skills to communicate their feelings, and may not have any friends—it is up to the professional caregivers to be responsible for assessing for suicidal behavior, not the manufacturers of the drugs. I have had many children who were on medications that are not approved for children under the age of 18—in which case it was the prescribing physician at fault.

However, parents need to take responsibility for learning about the medication their children are on, rather than simply assuming the drug they have been given is appropriate. A little common sense and responsibility goes a long way when dealing with a child requiring medication. Below is a list of medications recognized by the FDA as approved for children. Of note, there are no anti-anxiety medications approved for children under age 18.

Psychotropic Medications Approved for Children

	Trade Name	Generic Name	FDA Approved Age
Antipsychotics	Abilify	Aripiprazole	13–17 for schizophrenia, bipolar, and depression
	Haldol	Haloperidol	3 and older
	Orap	Pimozide	12 and older for Tourette's
	Risperdal	Risperidone	13 and older for schizophrenia; 10 and older for bipolar mania and mixed episodes; 5–16 for irritability associated w/autism
	Thioridazine	Thioridazine	2 and older
Antidepressants	Anafranil (tricyclic)	Clomipramine	10 and older (OCD only)
	Lexapro (SSRI)	Escitalopram	12–17 for major depressive disorder
	Luvox (SSRI)	Fluvoxamine	8 and older (OCD only)
	Prozac (SSRI)	Fluoxetine	8 and older
	Sinequan (tricyclic)	Doxepin	12 and older
	Tofranil (tricyclic)	Imipramine	6 and older (for bedwetting)
	Zoloft (SSRI)	Sertraline	6 and older (OCD only)

Mood Stabilizers	Depakote	Divalproex sodium	2 and older (seizures)
	Eskalith	Lithium carbonate	12 and older
	Lithium citrate	Lithium citrate	12 and older
	Lithobid	Lithium carbonate	12 and older
	Tegretol	Carbamazepine	Any age (seizures)
	Trileptal	Oxcarbazepine	4 and older
ADHD	Adderall	Amphetamine	3 and older
	Concerta	Methylphenidate	6 and older
	Dexoxyn	Methamphetamine	6 and older
	Dexedrine	Dextroamphetamine	3 and older
	Destrostate	Dextroamphetamine	3 and older
	Focalin (also XR)	Dexmethylphenidate	6 and older
	Metadate ER	Methylphenidate	6 and older
	Methylin	Methylphenidate	6 and older
	Ritalin (and SR)	Methylphenidate	6 and older
	Strattera	Atomoxetine	6 and older
	Vyvanse	Lisdexamfetamine dimesylate	6 and older

Source: National Institute of Mental Health
(www.nimh.nih/gov/health/publications/mental-health-medications)

In terms of ADHD medications, Strattera is the only non-stimulant medication currently available. In looking at the above list, based on the generic name, it appears that most of these drugs are by and large the same. That is because some of these medications are short-acting or long-acting, and some are in chewable or liquid form rather than solid pill form. There are also patches available.

So why give stimulants to a child who is already hyperactive? There is a misconception about Attention-Deficit/Hyperactivity Disorder. Children with ADHD are not hyperactive because their brain is causing them to be hyperactive—just the opposite. They are hyperactive because their brain is causing them to "zone out"—hence the inattention part of the disorder.

Children have a hard time focusing and keeping themselves mentally alert and awake, and so they do what they naturally do when they are sleepy and need to keep awake. They fidget. They pick up the first object they find and play with it. They become physically active, getting out of their seat and moving around, in order to wake their brain up. This is why when they are on their medication (a stimulant), they are calmer and able to focus. They no longer need to use their bodies to keep their brain active and alert, the medication is doing it for them.

Again, imagine yourself at a seminar for eight hours. You have just returned from lunch, and maybe you ate too much. Your stomach is digesting your meal, and you have already had to sit and listen to a lecture since 8:00AM. Now it is 2:00PM and all you want to do is fall asleep. The lecturer realizes he is losing his audience and asks everyone to stand up and stretch. Maybe you go get a coffee, even though you know caffeine is bad for you after lunch. You've had your caffeine and your stretch break, and suddenly, you are able to focus on what the speaker is saying again.

Children with ADHD do not know why they can't focus. They can't articulate it, and do not realize it is their brain. They think they are stupid. Everyone in the class seems able to complete their assignments, so it must be a problem with them, right? But they do not want to ask for help, because they know they can do the work. They just can't seem to concentrate! As adults, we learn coping skills to deal with how our bodies work. Asking a seven year-old to cope with something he thinks is a big shameful secret is a huge task.

As mentioned earlier in the chapter on statistics, there is an increase in the use of psychotropic medications among the autistic population from ages 6–12 and 13–17. As I speculated, this is probably a result of parents who previously were against medication finally resorting to medication when their child's hormones kicked in and their aggressive or self-injurious behaviors became unmanageable through other behavior modification methods. Most clients with neurological based psychiatric disorders which need medication management, and this includes pervasive developmental disorders, psychotic disorders, mood disorders, and obsessive-compulsive disorders, will see an increase in their behaviors when they reach adolescence.

If these clients are already on medication, their medications may need to be adjusted. Deonte, the psychotic student I discussed in an earlier chapter whose parents were reluctant to put him on medications and only did so after being told it was a condition for Deonte staying in school, continually had to have his medications increased as he grew. Males can put on between 10–20 pounds a year when they have growth spurts, and these changes have to be accommodated for with their medications. When Deonte came back to school one year after not attending summer school, his behavior had deteriorated back to ground zero—he had gained approximately four inches in height in one year and his delusions and physically aggressive behavior was back.

Brianna, the low-functioning autistic who was inappropriate for counseling, when she turned fourteen her mother stopped her medication and Brianna started taking her underwear off in class and flashing herself to any boy or male staff who was nearby, while mouthing sexually inappropriate comments.

Adolescence can be challenging enough when you have normal brain chemistry. When your brain is chemically imbalanced, all those raging hormones and emotions become even more difficult to manage. It is the difference between an "A" ticket ride and an "E" ticket ride. Only instead of warnings about people with neck injuries, back injuries, women who are pregnant, or people in poor health, there should be warning labels about people whose synapses are not firing properly.

I am personally all for homeopathic and eastern medicine. I do yoga, pilates, exercise regularly, and eat healthy. I have used herbal supplements and acupuncture. But I also know that where the brain is concerned, sometimes a good dose of Risperdal is the only answer. Asking a child to stop listening to the voices in his head by sheer will power is not only absurd, it is a danger to the client and others around him.

23

Nutrition and Psychiatric Disorders

While many of the parents of my clients are unwilling to put their children on medication, either due to fear of weight gain, or because they do not want their children on chemicals, it is ironic how many of these same kids who are not on medications are obese due to poor nutrition. When you examine the ingredients of the foods they are consuming, which mostly are prepackaged products targeted towards kids and/or lower socio-economic families, they contain man-made products and chemicals. So apparently, food coloring, food preservatives, and refined sugars in high quantities are acceptable, but a medication that will help stabilize a child's behavior and be more successful in school is out of the question.

The CDC reports that obesity among children ages 6–11 was 17% in 2006 and 17.6% among teens 12–19. Obesity is linked with bone and joint problems, sleep apnea, and poor self-esteem. I have worked with several obese children under the age of fifteen who were unable to climb the one flight of stairs up to my office without becoming out of breath. These children would frequently beg lunches from their peers and eat double or triple meals during the day, and during physical education sat, immobile in the shade, refusing to participate (students are required to do a minimum of ten laps around the quad).

In the previous chapter I discussed briefly how medications address serotonin, dopamine, and norepinephrine which are chemicals in the brain, which when a person has a certain psychiatric disorder they can have too much or too little of one or the other of any combination of these chemicals. These chemicals are also created as a result of the types of food we eat. Certain foods release certain chemicals. By changing a diet to healthier, balanced meals, a child can better regulate their mood.

Obviously diet alone will not alleviate manic phases of a bipolar disorder or eliminate depression—but it helps to address both. Many of my bipolar clients are overweight, and tend to diet by skipping meals in order to lose weight. This can cause the person to binge and overeat, resulting in too much sugar. People with bipolar disorders need to limit their sugar intake, eat healthy meals on a regular basis (to avoid crash and burn), and drink lots of water. People who are depressed often overeat (also known as comfort eating), which causes them to gain weight, which makes them feel more depressed, which starts the cycle over again. Additionally, because they are depressed, they often eat foods that are less than healthy such as chocolate or ice cream. Some studies suggest that too many sweets can actually increase depression (Fleming, 2007).

I will often discuss with my clients their eating habits, and I use the Overeaters Anonymous acronym, HALT, as a tool to help them. I will explain to them that when they reach for food at home, they need to think—am I Hungry, Angry, Lonely, or Tired? For my kids I also used the word Sad for Lonely and Bored for Tired. I tell them that if they are feeling any emotion other than hungry—they need to put the food down. If they are angry or sad, they need to write about their feelings in a journal. If they are lonely they need to call a friend. If they are bored or tired, they need to go for a walk. My adolescents especially love this—because it's empowering. They hate feeling like food has control over them.

Another exercise I do with my students is to bring out two different breakfast bars and show it to them. One is a popular brand sold in all grocery and discount stores. It features a fruit and has the word nutrition practically on the cover. The other is a less well-known brand that is sold predominantly in whole-food and organic stores, but the individual flavors tend to sound more like desserts. I'll ask them, based on the cover, which they think is healthier. They always pick the fruit-based bar. Then I turn them over and have them look at the nutrition facts and ask them what they see.

As they read the calories, calories from fat, dietary fiber, sugars, and protein, I write the information down on my indispensable whiteboard. The fruit-based bar will contain less calories, but about twelve grams of sugar to one gram of protein. Inevitably, when they read the other bar, they sound more surprised. There are more calories, but there is a higher percentage of fiber, and the grams of sugar and protein are the same. I'll ask them, "So what does that tell you?" Usually, they know that the one with the same ratio of protein to sugar is the better bar. Then I'll have them read the ingredients, and the first several ingredients of the fruit-based bar are various forms of sugar, while the other bar has sugar much further down the ingredient listing.

Then I explain to kids about advertising. I point out that just because a product says it has "fiber," "grains," or "vitamin" in the name, does not mean it doesn't also contain sugars and bad ingredients such as oils and fats. I teach them to look beyond the advertising and read the nutritional content and ingredients. I say, "I know you *are* smart, so you need to shop smart."

I especially do this with sodas. I'll tell then, "Hey, you keep buying those sodas. The owners of the companies put those ingredients in there so you'll get addicted and buy their product and make them rich. So keep putting money in their pocket so they can buy another summer home in the Bahamas. They don't care if you get diabetes, because they'll be snorkeling off their private yacht you helped them buy." This is probably the only area where reverse psychology actually works.

Kids don't like to be manipulated, but they like to feel empowered. If you teach them about nutrition and advertising, they like to feel that they can take control over their own health. I do not rely on the parents to do it. Sometimes my older kids will even have me write down all the bad ingredients so they can take the list home to go through their cabinets or show their parents.

The reality is, the food kids are getting in school is primarily sugar. When given a choice between a bottle of water or a soda, kids choose soda. When given a choice between fruit and nachos, they pick nachos. I've seen kids throw away enough milk, fruit and packets of vegetables to feed a village in China. But give them a bag of a processed snack food (that contains three servings) and they will eat the entire container in one sitting. Children do not understand about portions per packaging. They will tell me, "But a soda only has 110 calories." And then I ask them how many servings is on the bottle they are holding, and they say, "Oh. Two and a half." So then we do the math about how many calories are in the whole bottle they just drank, and how many grams of sugar that totals up to. Suddenly a glimmer of understanding begins to dawn in their eyes.

It is possible, with patience, to change kids eating habits. Several of my kids who want to lose weight, now purchase water from the school vending machines instead of soda. They used to buy the water marketed as sports drinks or vitamin drinks, until I educated them about the calories. I tell them if they want water with flavor, to simply buy regular water, and add a slice of lemon or crystal light.

While there is nothing I can do about the food they get at the school, I can help my students make better choices. Instead of eating the breakfast cereal provided by the school, which tends to consist of a high sugar to protein ratio, as well as a high sodium content, I suggest kids make hard-boiled eggs at home on the weekend (or have their parents do so), and they can bring an egg to school to eat in the morning or before they get on the bus. Hard-boiled eggs will last in the refrigerator for a couple of weeks. If their parents bring them to school, and they have time to eat breakfast beforehand, I suggest they have a grilled cheese sandwich at home with low-fat cheese on whole wheat bread (without the additional sugars added which is surprisingly hard to find).

Nutritional Information from Foods Targeted To Kids*

	Average Kids Cereal	Average Cookie	Average Chip Product	Packaged Fruit Snacks
Calories	220	410	388	83
Calories from Fat	4	167	208	10
Total Fat (g)	1	19	23	1
Saturated Fat (g)	0	6	4	1
Trans Fat (g)	0	1	0	0
Sodium (mg)	253	477	650	22
Total Carbs (g)	53	57	39	18
Dietary Fiber (g)	2	2	2	0
Sugars (g)	24	20	2	13
Protein (g)	3	5	5	0

*Products consisted of Frosted Flakes, Raisin Bran Crunch, Corn Pops; Mini Chips Ahoy, Ritz Bits, Teddy Grahams; Cheetos, Frito Lays, Doritos; Fruit 'n Yogurt, Yogo Bites, and Care Bears Fruit Snacks.

Among the overall kids products sampled, bagged chip products had the lowest sugars overall, but the highest percentage of calories from fat (54%) and the highest level of sodium (650 grams) on average. Among the cereals, the one with the healthiest sounding name, Raisin Bran Crunch, since it has fruit in it, was the worst offender in terms of calories from fat (12 calories) and the highest in sugars with 30 grams. This is because both sugar and high fructose corn syrup were listed three separate times on the ingredients. The best cereal in terms of sugars, Corn Pops, had only 18 grams of sugar, but only 1 gram of protein. Even so called healthy snacks—fruit based snacks, contain no protein, but lots of sugar. The solution—eat a piece of *real* fruit. Nutritionists advise parents to shop around the edges of a grocery store and stay away from aisles—packaged products are in the aisles—fresh produce, meats, and dairy products are on the edges.

With parents of younger kids, I can talk to them about making small changes in their child's diet. Eliminating calories in small places on a daily basis can add up over the week. For example, using fat-free mayonnaise, light dressing, switching to light juice drinks, diet sodas, and finding healthier cereal options where the sugar to protein ratio is closer. If you switch to light and fat-free products, you can eliminate 50 to 100 calories a day by those simple changes and the children won't notice the difference. Then parents can gradually introduce more dramatic changes such as eliminating the soda and packaged snacks completely.

When the kids choose a snack, I suggest pretzels, or I encourage them to save their fruit from their lunch as a snack. To get kids to eat vegetables, I suggest they eat carrots, broccoli, or celery dipped in light ranch dressing. The same thing can work with sliced apples or other fruit, if kids won't eat them alone. Light ranch dressing has significantly fewer calories than regular vegetable dip, and if they limit the amount that they put on the veggies (use mini carrots or small fruit slices so they only need one dip per slice) they will get their nutritional value of the fruit with minimal additional calories. Compared to pre-packaged fruit snacks which contain multiple levels of sugar, partially hydrogenated oil, and multiple types of food coloring, this is a healthy alternative. Eventually, the kids can transition to fruit and veggies without the dressing.

I teach the kids how to read the labels for all ingredients. I point out that the sodium in snacks makes them thirsty which makes them drink the sodas which has the sugar. Even juices, which are marketed as 100% juice, and as good for kids, are pumped full of more sugar than "energy" drinks. I had one student tell me his mom bought a particular brand of juice that was so good he drank the entire 64 ounces in one day. He of course was up all night, but as he put it, "It was the bomb!"

Below are some of the contents of some of the average drinks targeted to kids. While the average sports drink has only fifteen grams of sugar, most of the bottles stated they had between 2–4 servings per bottle. I think only once, in the last school year, have I ever seen a child throw a drink away unfinished.

Nutritional Information from Drinks Targeted To Kids and Teens* (Per Serving)

	Average Juice	Average Soda	Average Sports Drink	Average Energy Drink
Calories	140	112	60	105
Sodium (mg)	18	35	160	85
Sugars (g)	**33**	**30**	**15**	**26**
Protein (g)	0	0	0	0

*Products consisted of Juicy Juice, Motts, Welches; Mountain Dew, Pepsi, Root Beer, 7-Up, Sunkist; Gatorade, Powerade, Low-Cal Gatorade; Red Bull and AMP (energy drinks are not intended for children under 18, but teens drink them regularly).

Just as I can tell when my clients are not on their medication, I can tell when they are not eating healthy. I have certain kids—especially those with oppositional defiant disorders or learning disabilities, who I will not take to my office if they have just had candy or a soda. Their behavior will be considerably more hyperactive and defiant compared to if I get them first thing in the morning or after they have had their physical education. In other words, after a semi-balanced meal or physical exercise, they are well behaved, but put 20-30 grams of sugar in them, and they become a completely different person.

One teacher commented on one of my bipolar clients, that she could always tell when she was dieting, because her behavior in the afternoon after she ate lunch would become manic and defiant. She would start shouting obscenities, cursing out the staff, and threatening to AWOL. This would be because she would go without eating breakfast, and then her body, which is already chemically imbalanced, would have to deal with an influx of glucose from lunch after going without food since her dinner the night before.

A research study by the University of Michigan found a correlation between childhood obesity and behavioral problems. After following over 600 children for three years, they found that 21% of the children with behavior problems were overweight. Children with behavioral problems were also more likely to become overweight if they had been at a normal weight at the beginning of the study (the comfort eating problem). The CDC found among high school students who were overweight that 70% did not participate in PE daily, 35% watched three or more hours of television on an average school day, and 25% played video or computer games, or were on the computer for purposes not related to school for three hours or more on an average school day. In other words, nutrition is one part of the equation—a sedentary life style is another.

The problem is large and complicated. The government cutbacks in schools affect the food the students get (never mind that the government will ultimately have to pay the healthcare costs of people who get poor nutrition and have health complications as a result). Foster parents do not want to take the time to cook well-balanced meals. Biological parents are often struggling just to cope, and often have thirty years of bad nutritional habits themselves that they do not want to break.

However, if a child is depressed over his weight and brings it up, take the opportunity to discuss nutrition. Start the discussion early and often. When they complain about how they look, comment—gently—that at physical education you don't see them participate. Encourage them to walk if they don't want to play sports. Start with five laps. Then try ten. Build up to twenty. Kids often tell me they want to join a gym. I help them find things they can do at home that do not require money or effort on their parent's part. I tell them to walk up and down the stairs at their apartment building at home until they are out of breath. Then I tell them next time to add one more set. I tell them to buy things that are cheap that they can use outside, like a jump rope. When they say, "That's for kids," I tell them that Olympic athletes train using a jump rope. I give them a challenge to do 100 jumps without stopping and then see if they think it's for kids. Again, it is about empowering the child to take control over their own body and mind. It will help prepare them for when they are an adult and can no longer look to others to solve their problems for them.

24

Setting Treatment Goals

Every child who is placed in an SED setting comes with an IEP (Individualized Education Plan). This plan contains goals for the student for all areas of academics (math, reading, written language), behavior, vocational, and any additional services that they may receive such as speech, occupational therapy, or counseling. Of note, some students will have a social/emotional goal, which addresses their social development skills. This is usually applied to elementary school children or autistic children. A counseling goal is also usually referred to as social/emotional—so a student can have two social emotional goals. In order to avoid confusion, when I do my IEPs, I always title my pages "counseling" so that they do not get confused with social emotional when it comes time for report cards to be distributed.

Students transferring to a new school have a 30-day IEP to evaluate the goals from their previous school, and to see if these goals are still appropriate or need revision. All students have annual IEPs, which occur in the same month as their birthday (or the nearest available school month if their birthday is during a break). During an IEP, their teacher, parent or legal guardian (whoever has *educational rights*) is present, along with an IEP coordinator, school psychologist, and the various persons providing additional services. Sometimes a social worker from DCFS may be present, or a representative from an outside agency, as well. Depending on the age of the student, the student is also present. Every three years, a three-year IEP is held. This is the same as an annual meeting, only more comprehensive. The student will undergo intense psychological testing with the psychologist, be evaluated by a nurse and an audiologist, and an in-depth report will be written up.

Typical treatment goals for counseling usually deal with decreasing a negative behavior, increasing a positive behavior, or both.

For example, increasing frustration tolerance by staying on task longer, expressing anger and frustration in an appropriate manner (such as increase using their words and decrease physical acting out), increasing positive peer relations, increasing the use of conflict resolution skills, or taking responsibility for actions rather than blaming others.

Students get very nervous about their IEPs. For them, they are worse than their report cards. Everything about them is going to be discussed in detail. All their goals, whether they have obtained that goal, and what their new goal is. If any behavioral problems have occurred (suspensions, multiple incident reports), these will be addressed. As a therapist, I make certain to always attend an IEP, short of an unavoidable conflict. I check the IEP schedule months in advance, and ask that it be rearranged if there is a scheduling conflict and I cannot be present for a meeting, because my clients want me there for emotional support.

For example, one of my clients, Darrell, was unhappy in the class he had been in for the past two years. He felt he needed to be in the high school building with the older, more high-functioning students. He wanted it brought up in his IEP, which I suggested he do. So in session together, we role-played how he would ask for this class change to happen. We worked on his body language and his tone, since in the past in his IEPs Darrell tended to sit slumped forward with a scowl on his face, looking down at the table. We talked about making eye contact, and shaking hands with the staff when he entered and was introduced. Each week as his IEP approached, he asked me, "You're going to be at my IEP, aren't you Miss Jillian?" To which I replied, "Have I ever missed one of your IEPs?" Darrell knew I had not let him down in the past. As it was, the IEP coordinator brought up the class change herself, and my client did not have to ask for it, much to his relief. But at least he was well-rehearsed if the need had arisen.

When I set treatment goals for my students, I try to set a goal that is obtainable, and addresses only one problem area that they are struggling with at the time. Because the students that I work with are so dysfunctional, setting a complicated goal that addresses multiple issues is setting them up for failure, which increases their low self-esteem which in turn increases their acting out. Also, I conference with the teacher to see what behavior goal the teacher is setting so that the counseling goal and behavior goals are not identical; although it helps if they are complementary.

I have seen many students come into my school with a counseling goal that was completely inappropriate for their age or their level of functioning. For example, "student will learn to express his anger or frustration appropriately and will be able to verbally identify triggers for these feelings in 4 out of 5 trials." This was a goal for a child of seven years of age who was academically behind his peers and as a result had low self-esteem. It was unnecessary to add the second part about verbally identifying triggers. A simpler goal of expressing his anger and frustration appropriately (i.e., with words, not physical aggression) would have sufficed. A goal with multiple parts can be setting a child up for failure because they have to achieve all the parts of the goal. Furthermore, the goal has to be something the student can understand. If the child cannot understand their own goal, how can they be expected to achieve it?

One of my young psychotic students came in to our school with this goal—"student will demonstrate improved self-advocacy skills by appropriately verbalizing his thoughts, needs and feelings to school staff when frustrated, anxious, confronting peer conflicts, or struggling with difficult classroom tasks 3 out of 4 times." Whew! Aside from the fact that I'm certain these students do not understand what self-advocacy means, he now has to verbalize his thoughts, needs and feelings (while ignoring the voices in his head). That right there is a lot. Add on top of that, he has to do all of the above when he is frustrated, anxious, dealing with peer conflicts, or struggling with class work. Perhaps the therapist was paid by the word. Again, since this client was psychotic, there was no way he was going to achieve the above goal. This could be an appropriate goal for a non-psychotic, high-functioning older adolescent, although even then the goal tends to be verbose. A more simple goal would be something along the lines of expressing his needs appropriately when overwhelmed by external or internal stimuli (that way classroom distractions, emotional states, and hallucinations are all covered).

I work with my students in session for a month before their upcoming IEP together to set their goal. If a student is new to the school, and I have little to go on because I have only been able to see them in session three times, I use what is in their previous IEP, my discussions with the staff at school, and my observations of the client to go on. I should note here, that one month prior to my student's IEP I always do an observation of them both in their classroom and during physical education. The reason for this is because many of my students who have impulse control issues tend to be very polite and restrained while in session with me and there are no distractions or peer conflicts, but behave differently when in class. So while I talk to the teacher to get feedback on the student's behavior, I like to observe my client myself, because teachers often miss more subtle behaviors since their focus is split among twelve students.

For example, a young man was transferred to our school without any counseling goal. He was in a group home, and had been apparently removed from several group homes. None of the group homes wanted him. His IEP diagnosis was (surprise!) Autism. Freddy clearly did not have Autism. He was highly social, academically capable, there was a slight possibility of mental retardation—but only the slightest. At best, he could have been moderate-functioning Asperger's, but that is not in the same classification of Autism at all. His behavioral problems were clearly more oppositional defiant disorder.

Freddy and I were talking about his upcoming IEP, and what his counseling goal should be. He was unable to come up with one. So I asked him, "Why are you getting kicked out from your group home?"

"The other kids are bothering me. They're touching my stuff, and taking it without my permission."

"Hm," I say puzzled. "Well, that's what the other kids are doing. They would not remove you from the home for something the other kids are doing."

"But they've been there longer."

"That doesn't matter." I explain to Freddy that I worked in a group home for two years, and I know how group homes work. I ask, "What do you do when the kids touch your stuff?"

"I slam doors."

"Is that all?"

We eventually uncover that Freddy is easily provoked and damages property, uses profanity against staff and his peers, and is the only client on a point system at the home because of his failure to do his required chores, such as clean his room. He is also upset because one of the other clients at the group home is in his class at school, and he is easily provoked by this particular student.

Freddy states, "If he bothers me, I'm not going to be responsible for what I do."

"Oh, really? You aren't responsible?" I get up and go over to where Freddy is standing, and mimic puppet strings, "I don't see any strings. Nobody makes you act a certain way. The only person controlling your behavior is you. Not the staff at the group home. Not the kids in the group home. Not the kids in your class. You. You have control over how you react to others."

At this point, Freddy covers his head, and gives an exasperated sigh.

"Ah, ha," I say, getting out my pad of paper. "I've just figured out your counseling goal! Ignoring others." As I write this down, Freddy looks aggrieved.

"But I can't do that!"

"You can," I tell him, "You just choose not to." Freddy was already in high school, and we talked about the consequences of what will happen if he does not ignore others when he has a job. Will he quit every time he has to work with someone he does not like? Will he continue to run away from his problems? Or will he learn to find coping skills?

With a student like Freddy, who has an oppositional defiant disorder and blames everyone else for his behaviors, I could have set a more challenging goal such as "taking responsibility for his actions." Unfortunately, given that his environment was so unstable having moved from group home to group home, and constantly changing schools, I needed to make it as simple as possible to make a goal he might possibly achieve. The reason—it appeared—that he was getting kicked out of all these facilities was that he was easily provoked. He was not starting problems—he was sensitive to what others did and over-reacted. If I could get him to stay on task (doing his work at school, doing his chores at home) he would not be aware of what others were doing and would therefore not be able to be instigated into reacting. In other words, he needed to ignore his environment. Once we accomplished that goal, then we could step it up a notch and expect him to take responsibility for his actions, but it is easier to do that after his behaviors are more under control.

With some of my higher functioning students, I'll ask them whether they think they've achieved their goal prior to their IEP. Some of them will give me a very accurate answer. They'll say they've met the goal half-way, or that they've met it almost, but not quite. By working together with a student, you again help them feel empowered, so that it isn't you against them. Obviously, working with an older adolescent is different than working with a six year old. Although I have had precocious eight year-olds give me accurate feedback about their performance as well after I give them my patented look and ask "really? You think so?" if they tell me they've met their goal.

Some students will come up with their own goals, and do a surprisingly excellent job. They will tell me, "Oh, I really need to focus on my anger issues right now. It's getting me in trouble with the staff. It wasn't a problem before, and I don't know why it is now, but I want to deal with that so I don't have that problem when I get a job over the summer."

Part of the IEP is the student's Present Level of Performance (PLOP), which as a counselor I need to write up. This will discuss the student's strengths, needs, and the impact of their disability on their performance in school. I make certain to write my PLOP well ahead of the student's IEP, and I read this to my client before we have the actual meeting—assuming the child is old enough to attend the meeting. I do this so that when I discuss these issues in the meeting, they will not be surprised by anything I say, and will not feel betrayed.

Generally, what tends to happen is the clients focus on all the negative information. I'll ask them, "What did I say that was positive about you?" They usually will be unable to answer that. Or they will say, "Nothing!" So I reread the strengths area of the report to remind them of all the positive things I said about them.

Other times, kids will complain that the negative behaviors I mention in my report they don't do anymore. I remind them, gently, that this is an annual IEP, and it focuses on their behavior for *an entire school year*. Not just their behavior in the past two months. And then we can have a discussion on how it is important to act good all the time, not just the few months prior to an upcoming IEP (it is amazing how good students begin to behave when an IEP is scheduled).

For the most part, by including my clients in the process for the entire month prior to the IEP meeting, they feel like a participant. They don't feel betrayed or judged. And when the school psychologist asks them in the IEP if they disagree with anything I said, they invariably shake their head and reply, "No."

For a student in a non-public school setting to return to public school, they must achieve their behavior and counseling goals. This was the problem with my student, Andy, who usually met his academic goals, but was unable to achieve these two goals year after year. The reason for this is because public schools have SED classrooms that can accommodate different academic levels, but they are not equipped to handle behavioral and/or social emotional problems. They do not have the counseling staff, classroom aides, or enough staff to intercede if there are physical altercations on the campus.

So whenever I get a new student who wants to immediately leave my school, I tell them they have to have perfect attendance for six months, perfect behavior, and they have to meet their counseling and behavioral goals. They always ask me about their grades. I say their grades have to be passing, but the most important part is their behavior. Every time there is a physical altercation, the six month clock starts over again.

Of note, when an IEP takes place, it is an excellent opportunity to find out from the parents how your client is behaving at home. At my school, the counselors usually start off the IEP meeting. However, before I talk about my student's performance, I always ask the parent, "First I'd like to ask Mr. and Mrs. X how Johnny is doing at home. Do you have any specific concerns regarding his behavior?" That allows the parents to tell me what is going on at home first, and express any concerns they might have about what their child is telling them about problems they may have mentioned in school. Also, it tells me in advance if the parents and I are on the same page, or if they are going to be surprised by what my evaluation of their child is. Usually, when parents hear what I have to say about their child, their response is along the lines of "You've hit the nail on the head."

Then, once the parents are convinced that their child has not been able to manipulate me into thinking their child is a perfect angel, a barrage of issues at home will come up—such as how Johnny stole $200 from mom's wallet, or broke the window in the living room with a baseball bat, or is gang-banging and staying out all night tagging.

If you set the right tone with the parents in the initial IEP, the future relationship with them should go smoothly.

Furthermore, your client will be aware that they can't manipulate you into thinking things are great at home, and they only have problems at school, because you start off the meeting by letting the parents vent and have their say. Often, the parents, just like the clients, feel that their feelings and frustrations go unheard. Using reflective listening skills in an IEP can go a long way in forming a collaborative relationship with the parents.

25

Focus on the Positive

It is easy in this field to focus on the negative. The schools use charting systems to mark the child's behaviors on an hourly basis and track their behaviors across the year. Students get marks for being off-task, non-compliance, physical aggression, profanity and the like. So it becomes very easy to always focus on the negative, instead of looking for the positive when working with your clients. Sometimes, when you have a student with multiple diagnoses—mental retardation, emotional disturbance, and oppositional defiant disorder—finding the positive can be even more challenging.

Without trying to sound Pollyannaish, I try to praise my students for whatever positive actions they do as often as possible. For example, whenever a student holds open a door for me, I thank them, and follow up with, "good manners!" If a student is coloring and does a good job of staying inside the lines, I compliment them on that. If a student does not understand a word I use and asks me what I mean, I praise them for asking for help. The students (and staff) have enough negativity on the job on a daily basis, so I try to look for the good—even if it is the smallest thing possible. There will *always* be a chance to jump on a child's negative behavior—but if I pass up the opportunity to praise my client's good behavior—it may be weeks before it happens again. Praise motivates better than punishment—and children with low self-esteem like to be praised.

The reality is it makes my day go better as well if I can focus on the positive. It helps build a better therapeutic relationship with the children if I focus on their positive behaviors as well. I can always take the positive behaviors they exhibit when with me—for example, holding a door open or saying "please" and "thank you," and we can work on expanding these behaviors to their peers to improve social skills. Or staying focused while coloring can be expanded to staying on task in math. But if you start off by focusing on all the negatives—the poor social skills, the acting out during math, and the physical aggression—the kids are less likely to listen to you if all they hear is criticism. Imagine going into a performance review at work and everything your supervisor had to say about you was what you did wrong. Would you want to continue working at that job? Would you give 100% to the company? You would feel negatively about yourself and the company if all you heard were the areas that needed improvement. The same is true with these children. And it is easy to feel emotionally drained yourself as a professional, if all you are doing all day is focusing on the negative.

I had a child on my caseload who came to me at age seven. He was oppositional, and possibly on his way to a conduct disorder. His parents were already in the process of having him removed from the home if his behavior did not improve. In my first meeting with him, he was argumentative, lied to me, talked about stealing, did not follow my directions, did not listen, and told me he came to my school because the kids at his previous school made up lies about him. Within the first month it became apparent that Larry was unable to play games. He was unable to read at a first grade level, and he was unable to listen to the rules of the game in order to learn how to play them. It was not that he had ADHD; he just felt he knew more than me, and therefore he did not have to listen to adults.

So after several attempts, I gave Larry a "no game" rule. I told him for one month, until he proved he could show me good listening skills and the ability to follow directions, he could not play games. Larry was having a difficult time transitioning to our school. He had not made any friends in his class because his social skills were so poor. He liked to play the "victim." He would often make up lies about his peers touching, pushing, or shoving him, when no one was remotely near him. He blew any little incident out of proportion, and usually he was the instigator.

In session with Larry, the first week we had no games, he choose to work on stencils. He had a unique way of coloring in the stencils. He would go through every color in my box, doing the stencils in rainbow patterns, which gave the pictures an overall psychedelic effect. As he worked, I praised him frequently, saying, "Good job, staying on task," and "Wow, that's very creative. I've never seen anyone make a rainbow castle, before. Good job!" Larry practically beamed from all the praise. For one month, Larry did art work or puzzles and I praised him.

When we reintroduced games, we started with Candy Land, which required no reading or counting. He did well, and I praised him on doing a good job of not tantruming when he lost, and not boasting when he won. I explained to him that his classmates would not like him boasting when he won, and that when you finish a game, you shake hands and say, "Good game!" He was able to progress to Connect Four and Sorry. It took him awhile to figure out the multiple rules of Sorry, but I would continually praise him with, "Good job counting," and "Good job remembering the rules."

After approximately a year of this, Larry turned from being a nightmare to work with in session, to someone I looked forward to. He has stopped talking back and lying. When he had conflicts with peers he attempted to discuss them to the best of his abilities. Had his academics in class improved? Not really. Had his peer skills improved? No. But we were at a place where we were able to discuss it without him feeling defensive because a relationship had been established and he knew I was on his side.

The lower the functioning level of the child, the harder you may have to look to find things to praise. But there is always something there. One of my psychotic kids had an impulse control issue whenever we played games, and he would start swearing like a Tourette's victim. Every time he made a bad move or I made a good move, the word "fuck," or "damn" would come out of his mouth. In his case, I had to use both consequences and praise. Surprisingly, this was a behavior that only occurred in session; his teacher reported that it did not occur in class. I told my client that while we played Uno, every time he swore I would make him take an extra card. When we played Sorry, I told him that every time he swore I would send one of his men back ten spaces. That actually stopped the swearing. However, I could see that when my client got upset, he would make a jerking movement, and I could see he wanted to swear, but didn't. Then I was able to use praise, and say, "Good job not swearing!" It got to the point where he would sometimes make a squeaking noise, then ask, "Did I swear?" checking in with me. It took six months to extinguish the swearing, and he continues to check in if he is upset, but he has not sworn in over two years. I continue to praise him for it.

One of my kids with a conduct disorder in the group home, was in the middle of talking to me and he said, "I don't want to talk about that anymore. If I do, I'm just going to do something I'm going to regret. I don't want to hurt you." So I praised him for recognizing that his behavior was about to get out of control and expressing it verbally rather than acting on it physically.

Kids with low self-esteem tend to feel badly about themselves. Hearing negative things increases their low self-esteem. Any praise will help increase their self-esteem, and encourage them to take risks. Maybe then they will gain the confidence to try to read, or do math, or play a sport during physical education. But without any self-esteem, they will shut down and try nothing academic related, because they are programmed to think they are dumb, stupid, or retarded. Some of the positive reinforcers I use are:

- ✓ Good job staying on task.
- ✓ Good job putting away the toys.
- ✓ I appreciate you sharing your feelings with me.
- ✓ Good job expressing your anger appropriately.
- ✓ Good job being honest.
- ✓ Good job ignoring others.
- ✓ Good job asking for help.
- ✓ Good job problem solving.
- ✓ Thank you for being so polite.
- ✓ Thank you for asking so nicely.
- ✓ Good job using your words.
- ✓ Good job walking down the stairs properly.
- ✓ Good job following directions.

A one-to-one that I supervised, was assigned to Gabriel, who had ADHD, social phobia, and obsessive-compulsive disorder, was also a hair-puller. Gabriel often verbalized his anxiety when he was about to pull someone's hair, and so we were frequently praising him for verbalizing his feelings, because it allowed his aide to follow through and ask him if he wanted his hands held together. Gabriel felt anxious because he was afraid he might hurt someone, and would say, "I'm afraid of your hair," which meant he was afraid he might pull the person's hair. We also frequently praised him for keeping his hands to himself. He liked having his hands pressed together and squeezed, and would ask for that, and we would praise him for requesting this action since these requests usually indicated an increase in his anxiety which preceded an impulsive gesture such as grabbing, poking, or pulling. Again, the more severe the behavioral problems, the lower the expectations of what you will be praising. These are not students where you are going to be congratulating them on writing a poem, or completing a science project.

One morning I went to get a student with mental retardation, and found she was on LOP for not having done her homework. It turned out that Karen had not done her homework because her mother had told her not to—her mother felt the homework was too hard. So Karen had been put in a double-bind situation—either disobey her teacher or disobey her mother. Normally I cannot take a student on LOP, but the teacher and I agreed the situation needed to be sorted out, so I took Karen to my office and called her mother. I explained to the mother how her instructions to her daughter had gotten her in trouble in school. The mother basically had been overwhelmed at home and did not have the time to help her daughter.

I spent Karen's session with her helping her complete her homework. Karen was capable of doing the homework assignment— she just needed someone to read it to her. Her cognitive skills were strong enough to answer the questions for the assignment, but her literacy skills were too low to read the handout. However, Karen was already frustrated by having gotten in trouble, and was in a bad mood.

We worked on calming down, then I read her the assignment. She answered the questions, and each time she answered a question I praised her with, "Good job listening to the material." Like young Larry, she practically beamed.

She stated dismissively, "It's easy. I just can't read it!" We finished the assignment quickly, and I praised her for staying on task and finishing her work. When I returned her to her class her mood was completely turned around from sulking and angry to smiling and upbeat.

With higher-functioning clients, using verbal praise repeatedly can work, or can backfire, causing them to get angry, stating, "Don't treat me like I'm special ed!" One of my older kids with Asperger's was seriously trying to work on improving his social skills and to determine why the other kids in his class teased him and did not approach him. So in session one day I mirrored for Paul his body language. This was a student who tended to look angry and pissed off all the time, with a frown on his face. I mirrored his facial expression, and asked him, "Would you approach someone on the yard who looked like this and start up a conversation?" The other problem with Paul was that he walked rapidly, and looked like a Neanderthal man walking across the yard—head bent down, arms dangling at his side. Aside from looking comical and making him a target for teasing, it was hard to catch up with him—even if a student wanted to approach him to talk, he would be gone before they reached him. So in the privacy of the office building, I demonstrated Paul's walk. He insisted he did not walk that way, but then later was convinced he did when his teacher confirmed this. Unfortunately, by looking at the ground when he walked rapidly, in addition to scowling, Paul came across as completely unapproachable.

Paul and I worked on how to walk properly, and whenever I saw him in the yard walking properly, I gave him a hand signal—a closed fist with thumb and pinky in the air, rapidly waved. Not as obvious as a thumbs up, but Paul knew what it meant. Within a few months, Paul's strange mannerisms were extinguished, and he was spending more time on the playground talking with his peers instead of being isolated. I actually started to see him smiling.

I use a lot of similar positive hand gestures with my older, higher-functioning clients on the yard when I see them engaged in positive activities. For example, when overweight students participate in physical education instead of sitting, they see my thumbs up, without me calling attention to them by saying anything verbally. That way they know that I am aware that they are making an effort, but there is no embarrassment by my verbally calling attention to their actions. More often than not, the students do not care about the embarrassment, and will come to me in session, boasting, "Did you see me playing volleyball? I'm trying to lose weight. I've exercised three times this week." As I said, for the most part, these kids are desperate for acknowledgement of their achievements, no matter how small they may be.

26

Working the System

I remember the first time I had to file a child abuse report. I was a trainee at a substandard agency where I was told, "Here's the paperwork, make the call," and I was left on my own. While I was filling out the paperwork, several interns came in to the conference room where I was working and commented, "Oh, making your first report. That's tough," and out they walked. No advice. No support. Afterwards the parent was furious with me. My supervisor had abandoned me. And I felt as though I had betrayed my client's trust.

The next agency I worked at had a different way of handling child abuse reporting among trainees and interns. You were given a "mentor" to walk you through the process. Someone who had made several reports basically held your hand as you filled out the paperwork, called the parent to advise them you were filing a report, coached you in what to say, and sat with you while you called the Department of Children and Family Services. Whether your first experience with filing a report is good or bad, one thing is for certain when working with this population—you will eventually file more reports than you can recall.

There are a lot of complaints in Los Angeles about the Department of Children and Family Services. I have to say this in its defense—no doubt the concept is good in smaller towns or cities. But in larger towns like LA, the system only works about half the time. However, it helps if you know how to make the system work for you.

First, let's examine why the system does not work. The sheer numbers. In a city with millions of inhabitants, and we have already seen that California is number one in foster care—there are too many kids who fit the profile for potential abuse. Secondly, there are too many families from different cultures where their ideas of child-rearing do not fit the criteria of the orthodox "American" concept of what is acceptable discipline. As a result, what might be acceptable in Persia or China, is considered abuse in America, but unfortunately, families immigrating to California do not get a pamphlet entitled "How to avoid being reported to DCFS." And finally, there are too many manipulative teens from dysfunctional families who abuse the system. I have had clients who will tell me they have called the police on their parents, hoping to get placed into an "adoptive" home.

These clients who want to get "adopted" are typically adolescents with drug abuse problems, truancy issues, multiple psychiatric diagnoses, and physically aggressive tendencies. I look them in the face when they tell me they want to get adopted, and I tell them, "Honey, there is no chance in hell of you getting adopted. People in California are adopting little baby girls from China and Romania rather than adopting kids from California. There is *no way* a black adolescent with a rap sheet as long as yours is going to get adopted. Here's where they are going to send you if you keep calling DCFS—a residential facility, or if you are lucky, a group home. Do you want to live with five other girls that you don't know and probably won't get along with any better than your own family?"

So for every thousand calls DCFS gets that is a legitimate call of abuse from a mandated reporter (teacher, therapist, doctor—anyone who comes in contact with children in their profession), they get hundreds of calls from manipulative kids who just had a fight with their parent who refused to buy them a cell phone or I-Pod and they decide to report their parent to the system in order to get a "better" family. Unfortunately, the department has to investigate these allegations just as they have to investigate the legitimate ones. So the department is spread too thin, and often the social workers are spending too little time on the legitimate cases because of all the time being wasted on cases based on no merit.

Here is an example of one such case. I had a girl on my case load, Leticia, who had major truancy issues. She used drugs, rarely came to school, and when she did, she was always getting into physical altercations or being sexually inappropriate. She had a terrible relationship with her mother, who never attended her IEPs. Every time a school break came up, we discussed coping strategies. For a year, every time Leticia returned to school after the break, I would ask her, "How did things go?" Her answer was always the same—great! She would tell me about how she and her mom went shopping, bought clothes, had fun.

Then after one break, Leticia came to my office and said, "If my social worker files a report to DCFS and they don't do anything, is there anything you can do?" I asked her what she was talking about, and Leticia told me the following tale. Apparently for the last six months her mother had been hitting her with a stick, her step-father had been physically kicking her in the head while her mother stood by and watched, and her mother was not feeding her and there was no food in the house (Leticia was overweight and always had been). Then the story got even more dramatic, and sexual abuse started coming into it.

I stared at Leticia, who was calmly building a puzzle in my office. "Wow. That's quite a tale. I don't know what to think. After every break I've asked you how things have been with you at home, and you've told me a different story. Now you tell me this has been going on for six months. So either you are lying to me now, or you've been lying to me for the past year, which is it?" Leticia insisted she was telling the truth. I told her I did not believe her, but I would file a report.

However, when I filed the report, I did so with Leticia in my office, so she could hear exactly what I said over the phone, and I reported it exactly as I presented it above. I told the worker over the phone that my client was either lying now, or was lying before. I told the worker that a report had theoretically been filed and the case dismissed, and that I personally suspected that over the break my client had a fight with her mother and was retaliating. One thing that I have learned is that DCFS usually appreciates any insight or additional information you can give them—after all, you have been working with the client for years—they have to go investigate and make a decision within an hour or less based on limited information.

Why did I make the call with my client in the room? So that she would know exactly what I said and not feel betrayed, or be able to come back to me and accuse me of saying something I did not say. Suffice to say, even though I told her flat out that I did not believe her, since she was clearly lying either now or in the past, she was not upset with me either then, or at any time later in the future (or at least, no more so than usual).

When a client is clearly unstable and manipulating, I do not make a report. I do not have to. The law states that I have to make a report if

I *suspect* abuse. A case in point was with a girl who had a conduct disorder and was a compulsive liar. She had physically assaulted several staff in the school, resulting in their having to be hospitalized. One morning she asked to see me first thing.

In my office, she claimed her father had punched her in the face the day before after church. I verified what time the alleged assault took place (less than 24 hours ago) and where he had hit her. I saw no bruise, no sign of even the slightest swelling. She was a slight girl in stature, and if a full grown man had punched her, there would have been physical evidence. I asked her what she wanted to have happen. She said, "I want to go live someplace else." I did not file a report.

The one thing that tends not to get reported enough is emotional abuse. While working in Watts, I had a boy on my caseload who was mentally retarded and lived in foster care. He was normally smiling and outgoing, but after a period of time, he started to lose weight, was unable to concentrate, and showed all the classic signs of depression. His teacher spoke to me, concerned, and told me that he had told her how his foster mom often threatened to "send him away" if he misbehaved. I discussed it with him in session, and learned that he was being emotionally abused in the sense that the foster parents were telling this child things that cognitively he could not sort out. He was literally afraid to do anything for fear of "being sent away." As a result of his fears, he was not eating, he was listless, and depressed to the point of suicidal ideation.

I had another case where a seven year-old girl repeatedly brought up in session the issue of her fears related to "going to hell." I asked her why she thought that was going to happen, and she told me her foster parents kept telling her that if she was bad, she would go to jail and burn in hell. After further questioning, it appeared that the foster parents were born-again Christians, and were trying to undue any possible future behavior problems the child might exhibit by using the fear of God. We discussed her fears at length, and I reassured her that she could not go to jail—she could not even go to juvenile hall. I then filed a report of emotional abuse, and when I spoke to the social worker I discovered that an investigation was already in process through another mandated reporter.

Sometimes emotional abuse can cause more damage than physical abuse because there are no telltale scars. The signs of abuse are more subtle, and if you do not know what to look for, you can miss it. Fortunately, the newer DCFS reports now have a box to check if the victim of abuse is developmentally disabled, which gives them a higher priority rating as being "at risk" since they do not always have the cognitive capacity to ask for help. The older forms did not have this category on the paperwork.

The case with Ahmed, my young student who died as a result of abuse not withstanding, I have had a fairly good response from DCFS. And it is possible to file reports for things you would not think are reportable, but get results—such as the time that Ahmed was overmedicated and there was no way to communicate with his parents.

I had a girl on my caseload in Watts who originally attended school fairly regularly. Angie was always an active participant in counseling, but I got the perception that her mother was either bipolar or doing drugs. After some time, Angie started missing school—a lot of school. But her mother was smart enough to know a truancy report could not be filed until Angie missed eleven consecutive days of school. So Angie would miss nine days, come in one day, then miss ten days, and so on.

When I saw Angie, I would ask her what she was doing at home when not in school. She was going to amusement parks, her mother was taking her shopping, to the movies, and to the beach. Angie would come in with manicures and pedicures, sporting a new hair style, new clothes—in other words, it sounded like her mom was on a manic shopping spree and was confusing her daughter for a friend and taking her along for a ride.

It is a law that children under the age of eighteen attend school. I went into the attendance records, and counted how many days Angie had attended during the semester, then filed a child abuse report for neglect. Her mother was neglecting Angie's education.

The Department of Children and Family Services have several ways to classify a case in terms of response time. Urgent means they respond in four hours. Or they can respond in 24 hours, 72-hours, or a week. They marked this call urgent. Why? Because Angie had a little three year-old brother, and there was concern over who was taking care of the toddler while all this crazy behavior was going on in the home. The next day, Angie was at school. I let Angie know that it was I who had made the report, and she was not upset with me. If you have a strong enough relationship with your client, and they know that you care about them, the therapeutic relationship will survive a child abuse report. If you make a report and the child does not speak to you afterwards, chances are the bond was not there to begin with. As I said, I have an honesty policy with my clients, and most of the time it works in my favor.

When making a report to DCFS, it is necessary to have all the information on the family available to you. The names and ages of all other siblings in the home. All phone numbers and addresses of both parents or legal guardians. Check with the administrative office of your school to assure the information you have in your file is the most current. Have the student's intake file pulled for information that might be missing, such as sibling information. If you make a call to DCFS and do not have the information on hand, or waste their time while on the phone looking it up, they will get annoyed with you. Be professional and have the facts before you call. If you can fill out the form before you call, you will be certain to have all the information they will ask for prepared. When you are done you can send off the form without having to worry about meeting the legal deadline. Often times, the presence of small children in the home can make or break the decision for the worker to accept the case. If you get a worker who insists that you do not have a case and you are certain that you do; hang up and call back. Chances are the next worker you speak to will be more sympathetic. But I have only had one case denied, and that was because there was no current address or contact information (the child had disappeared). It truly is how you market the pitch.

With Billy, my physically abusive young boy who assaulted his mother in the car while his father passively watched, I knew for a fact that there was a toddler in the home. I also let the worker know that in his play Billy had recurrent themes of death and killing family members. As a result of this information, DCFS responded promptly to my call. If a child is physically abusive to a parent while one passively watches, will they stand by while the child abuses the sibling? DCFS cannot take that chance.

Once a client has been assigned a social worker, or the Department of Children and Family Services deems that outside services are required and the client has a therapist or case manager from an outside agency, I try to always notify that person whenever my client has an upcoming IEP. If I catch an unfamiliar person on campus with one of my clients, I will go up to them and introduce myself, and ask who they are. I will ask if they have a business card, and staple their card to the inside of my client's file for reference purposes. That way, hopefully these outside people will attend the child's IEP. They rarely attend due to limited time and/or scheduling conflicts, but again, I have done the best that I can to make them informed. On the rare occasions when these outside representatives do attend, they can be a wealth of information into what is really going on inside the child's home and what interventions have been tried with and without success.

On the flip side of the coin is the PET (Psychiatric Emergency Team) division. Again, as with DCFS, I am sure this concept is an excellent one in say—Vermont or Maine, where there are only 600 children in foster care. However, in California, the response time of the PET team is so delayed, I have never once had a child detained on the first call. For example, while working in the group home, a young man barricaded himself in his room and hurt himself superficially. He was out of control, screaming he wanted to kill himself and everyone else. By the time the PET workers arrived, and the staff had physically removed the door to his room to get him out, it was six hours later and he was calm. So the PET team determined—because he was calm—he was not a danger to himself or others and did not put him on a hold.

A week later, the boy again barricaded himself in his room, this time shoving his hand through the window and using the glass to cut his arm one night. The PET team was called, but the group home was told it would be the following morning before they could transport him to the hospital. When the first incident occurred, the PET workers spoke to me on the phone, but despite my history with this patient, they refused to remove him. If they had listened to me, he would not have had the opportunity to harm himself later.

The same was true of the boy at the group home who had to be transferred to another house because he broke a window and attacked a staff (this was after I had the PET division called, and again they did not take him away). Unfortunately, as best as I can tell, the criteria for taking someone to a hospital for the PET division is that the person has to be holding a gun to their head—in which case, I would be calling the police, not the PET team.

Which leads me to my next point. Now, whenever I have a student who is suicidal or a danger to others at any place that I work—I call the police directly. I have never called the PET team since working at the group home (that was their policy). Fortunately, my school is more than willing to call the police, because one, the PET division will not arrive before the student leaves, and it is not good to put an unstable child on a school bus that may take two hours to reach their home. Two, there is no guarantee that the PET division will get a coherent story from the family, and after the school is closed they can not reach the staff at school to find out the situation.

As I said, the system can work. But it helps if you know how to work the system.

27

Dealing with Parents

One of the benefits of working with this population, especially in group homes and schools, is that you rarely have to deal with the parents. Part of the reason, if you haven't already ascertained this, is because most of the clients' biological parents do not have legal rights. Foster parents rarely have educational rights, so there is a representative from the Department of Mental Health in charge of their educational needs. And as I have reiterated frequently, foster parents do not bother to call the school unless absolutely necessary, because they usually do only what they are paid to do—provide for the child's basic food, clothing, and shelter.

On the other hand, when you do have to deal with the parents, they are usually more dysfunctional than the children. In psychological parlance, there is a term, "identified patient," which is when the parents come to private practice and say, "There's something wrong with my child—fix him." That person is the identified patient. In reality, however, the child is acting normally to the environment around him. Let's say the child is eight, and his parents are arguing, and there is the possibility of a divorce. The mother is depressed and often locks herself in her bedroom crying, and the father works late to avoid being at home dealing with the marital discord. So as a result, the eight year-old starts having problems concentrating at school, he has nightmares, maybe has outbursts and fights with his peers and siblings. However, his siblings do not seem to be acting out because they have a better understanding of the family dynamics and realize it is not about them, and they have the cognitive capacity as older children to discuss the problems with their peers at school. In other words, they have emotional resources not available to the eight year-old.

When working in a school, the child is often the identified patient—sometimes rightly, but often the parents also have issues which they are in denial about. Many times I will recommend family therapy to my clients at the school, and the parent will say indignantly, "I don't need no family therapy—Billy's the one with the problem. Not me!" To which I will reply, "Well, then maybe you would like some individual therapy. Coping with someone as difficult as Billy must be hard on you. Perhaps you would like someone to talk to about how this has affected you." Even then, the parents will deny that *they* have a problem.

On the other end of the spectrum are the parents who deny that their child has a problem, and they simply can't understand why their child gets expelled repeatedly from every school they attend. It's the teachers. They aren't compassionate enough. They do not have the right training. It's the wrong school. They do not run the right program for their child. No one understands their child. There is a saying about vacationing in a river in Egypt—some of the parents I have worked with live on the river Nile. Here are just a few of the more extreme examples.

Tommy had Asperger's and terrible hygiene. He never brushed his teeth, and claimed he did not know where his toothbrush was. His hair was unkempt and uncut. His clothes were two sizes too small for him, and smelled. Every time I picked Tommy up I would ask him, "Did you shower last night?" and he would always reply that he forgot. Tommy initially did his work in class, but would tantrum when asked to do something he did not want to do. He would spit, bite, kick anyone who came into his range, and would basically sit on the floor in the fetal position like a wild animal screaming if anyone came near him. I guess I should mention Tommy was in middle school. One time I happened to be in the building and heard his yells, and watched one of his tantrums. He was red in the face, and kept screaming, "I'm king of the universe! You can't tell me what to do!"

Tommy's parents had explained to us in his first IEP that when Tommy was little, they were so in awe of his academic brilliance, that they let him do and have whatever he wanted. In other words, they never told Tommy "no," and so he never went through the normal toddler stage where he learned boundaries or discipline. Now Tommy weighed over 130 pounds and they were unable to make him do things that a normal seven year old does with minimal prompting (such as combing hair, going to bed, teeth-brushing).

Tommy had no friends due to his poor hygiene. It was discussed with his parents who brought him to school on a monthly, if not weekly basis. Nothing changed. Every IEP it was brought up. Tommy's parents kept telling the school we just did not understand their brilliant child.

Meanwhile, Tommy stopped going to bed at night, and was up doing whatever he wanted. So he started sleeping at school during the day, and failing his classes because he was not doing his work. He may have been academically ahead of his peers when he was little, but by the end of the time I worked with him he was behind his classmates because he never did his work.

Tommy had come to our school because his parents pulled him out of his previous school due to a staff member filing a child abuse report on them. They were insulted that someone could think they were abusing their child. Just because they neglected his medical needs and personal hygiene, and his behavior was out of control. The mother also felt the neighbors were against them. All the parents in the neighborhood would not let Tommy come over because he always provoked the neighborhood children, and the parents accused her of being a bad parent. Tommy's mother just did not understand it. Why was everyone against them?

One day Tommy came to me and he showed me his thumb, which was green with puss. He said he had hurt it weeks ago, and it was hurting. I asked if he had been to the doctor, and he said no. I asked if his parents knew about the injury and he said yes. I filed a child abuse report. The next day, Tommy was not at school because he was at the doctor. The next time there was an IEP, Tommy's mother was furious that the school had filed an abuse report instead of calling her. The mother said they were looking for another school for Tommy (she was unable to find a school that would accept him due to his physically aggressive behavior).

Eventually, Tommy was sleeping so much in class, that I filed another abuse report. The interesting thing was that Tommy kept telling me about things going on in his home, almost as though he wanted me to make the reports. When kids are so out of control, they want someone to come in and give them structure. Just as the kids prefer the structure of school during the school year, and dread the holiday breaks—kids whose families are in serious dysfunction, are practically crying out for help.

When the DCFS worker came out to discuss the case with me, he observed Tommy in school and talked to him. He then went to Tommy's home that day. The worker came back to the school and reported that in all his years of working for DCFS, he had never seen a child's bedroom so unsafe and disgusting in his career. He gave the family one week to resolve the problem or Tommy would be removed from the home. Tommy's parents continued to insist that there was nothing wrong with their child, and they continued to look for another school. The problem with Tommy's behavior was the school—it was not their child, and it certainly had nothing to do with their parenting!

Another parent who was difficult to deal with—although fortunately my contact with this father was limited, was the father of a young man I worked with who was higher functioning. The father, who was unemployed and probably living off welfare or disability, was also an alcoholic. However, James was in high school and worked hard in counseling. His younger brother was also in school, and not doing as well. The father never attended IEPs, although sometimes he would telephone in for a phone conference during the meetings, sounding as though he were drinking while the meeting was going on. He never heeded any of our treatment recommendations (family therapy, after-school tutoring, psychiatric evaluations), but always had an opinion about our school program.

One day, and I honestly can not remember what the circumstances were, the father showed up at school for a meeting with the principal and both his sons. Myself and the counselor for the other son were present. During the meeting, the father started going off, ranting and raving about how our school was responsible for his youngest son's behavior problems, and that he wanted to pull both his sons out of the mental health program. At some point, I interrupted the father and pointed out that he could not blame the school for his son's problems, when he had never once attended an IEP, had not once followed up on any of our treatment recommendations, which I then spelled out. The father stared at me, and shouted, "Who is this woman? I want her out of here! I won't continue this meeting with her present!"

I sat my ground, and my client stared at the floor, embarrassed. Once it became clear that the parent could not intimidate me the meeting continued. Ironically, about a year later the parent would realize that I was helping his son, and would be civil to me, actually asking if it was possible for me to see his son twice a week. Unfortunately, this is typical of the type of abuse you have to take with parents of dysfunctional kids. Often times, when you see the parents, the origins of the children's emotional problems become crystal clear.

One such example was Sara's mother. I had worked with Sara for four years. Sara's mother had a myriad of problems. She was unable to maintain a job for more than two months, and the family was often homeless. They lived out of motels, homeless shelters, with friends, and who knows who else. Due to Sara's shame, she often did not want to talk about the situation. As a result, Sara had difficulty attaching to people. She couldn't make friends because she never knew how long she would be in one place. They could be there a few weeks or a year. Sara had a lot of guilt, anger, depression, and grief over the inability to protect her younger siblings from this lifestyle.

Sara rarely talked in counseling, but she came every week. I was the one person in her life who was dependable—who did not change year to year. Her teachers and aides changed every year (sometimes more frequently), but I was there for her since she came to the school.

Sara's mother started giving Sara lots of things to make up for what was missing in their life. Big screen televisions, cell phones, computers. Sara's room was an electronics haven. As a result, Sara started missing school. Why come to school when you have all these expensive toys at home? Academically, Sara could have returned to public school years ago. Her grades were smack on. Her behavior at school was perfect. Her teacher approached me suggesting she return to public school, and I agreed. There was one problem, her attendance. So in counseling, for a year, I kept telling Sara that she needed to attend school if she wanted to return to public school. I even went so far as to threaten her with the option of residential placement to ensure she went to school.

Sara's IEP came up, and her mother, who I had spoken to for the past three years at previous IEPs came in and sat down. She turned to me and said, "Are you Sara's therapist? Good, I've been wanting to talk to you!" Sara's mother then went on to accuse me of causing Sara to physically assault her in her home, to cause Sara's behavior to go out of control, and to turn her life into a living hell. I sat there, stunned. How had I done this?

Apparently, Sara's mother was under the impression I had told Sara she could return to public school. I asked Sara's mother, "If you had a concern about what was going on in your daughter's counseling session, why didn't you call me?" She said nothing, and stared straight ahead. I asked the question again. She said, "I want another counselor for my daughter. It's my right."

I turned to face her and said calmly, "If you want another counselor, I'm not going to stop you. I would like you to answer my question. Why did you wait until this meeting to bring the issue up? If you have a concern about your daughter, you should call the school to discuss it, not keep it bottled up inside you for months on end." Her mother continued to stare straight ahead, and kept repeating that she wanted another counselor. I attempted to explain that I had never told Sara she could return to public school. I even said that I had told her she would have to go to residential school if she didn't start attending school more regularly.

However, I now saw why Sara didn't communicate her feelings very well. She had learned it from her mother. The sad part was that this poor child who was used to having to give things up, was having the one thing she had known for four years being taken away from her—her therapist, all because her mother had poor communication skills. It also, apparently, never occurred to the mother that her daughter, who apparently was capable of assaulting her, might have been lying.

The worst parent I have ever had to deal with by far was Mrs. X. Her daughter, Brianna, was a low-functioning autistic. Brianna's behavior would get so out of control, that Mrs. X would come and pick her up from school early. Apparently, Brianna's parents both lived off disability or welfare. It was a well-known fact that neither worked, although they always had plenty of money to take Brianna to amusement parks and to buy her whatever she wanted. I had worked with Brianna for almost four years and felt Brianna was not making any progress.

There were two reasons for this. One, Brianna was too low functioning to benefit from counseling. Two, and more importantly, Brianna's parents were sabotaging all the rules of the school's programs. When Brianna misbehaved they did not consequence her, they rewarded her. Also, their behavior was so low functioning, that they were not capable of modeling appropriate behavior for her.

Brianna lived on junk food and soda. She was supposed to be on medication but was not. So she was often physically violent. But her parents were unwilling to give her the medication. But every time they came to the school to pick her up they gave her chips, ice cream, soda, or cake. As mentioned in the chapter on nutrition, sugar will increase bad behaviors. I do not recall ever seeing Brianna eat a single piece of fruit or vegetable. She never drank milk or juice. She only drank water and soda.

Brianna was hypersexual, and would fondle her breasts and rub up against people inappropriately. Her parents would come and pick her up, and then staff would see them standing outside on the street, having ice cream. Brianna was bad, so let's eat an ice cream to celebrate.

I felt that Brianna being in therapy was a waste of time. She would have been better off spending more time in occupational therapy where she could have learned fine motor and gross motor skills. So one day I told Mrs. X when she came to the school to pick her daughter up that I was going to recommend in Brianna's upcoming IEP that her counseling be terminated. Mrs. X went ballistic. I explained that her time would be better spent on occupational therapy or speech.

Mrs. X then went on to tell me that I had a bad personality, I would never get a raise or job anywhere else in my life. I would never get a promotion. I was a bad person, and that I was going to be miserable for the rest of my life. Quite honestly, it was all I could do to keep from laughing in her face. I wanted to tell her to get down on the ground and kiss my feet since she was living off my taxes.

What I did say was, "You're entitled to your opinion, and I'm entitled to mine. If you want Brianna to have another therapist, go to the administration office and request one." And I turned on my heels and walked away before I said something unprofessional. When you are dealing with crazy parents, all you can do is let them vent, then walk away. Fortunately, my many years in the entertainment industry prepared me for dealing with unstable people who threw bigger tantrums than any parent I have had to deal with—and they got paid a lot more money.

Brianna has a new therapist, and she continues to spit at people, give everyone the finger, says "fuck you," is sexually inappropriate, and gets to go home early half the time because her parents reward her bad behavior by picking her up early and giving her sugar. But at least I do not have to deal with Mrs. X anymore.

Not all parents are dysfunctional. There are parents, and some guardians (usually relatives who have legal guardianship), who make an effort to do what is in the best interest in their child. These parents will actually go the extra mile to help their child. These are the children who generally end up returning to public school, or graduate from an SED school with their diploma to go on to college.

One such student was a young boy named Jason. He was around seven when he came to my school, and he had been expelled from his public school for physically aggressive behavior. Jason quickly adapted to the school's program, but he had a lot of self-esteem issues which we worked on in counseling. His mother was an active participant, and called the school frequently, conferencing with the teacher, and attending meetings. She would call parent-teacher meetings when she had concerns in between IEPs.

One of the areas holding Jason back was his peer skills. He tended to have an authoritative, almost bossy attitude, without quite crossing the line of being an actual bully. It was more that Jason felt he was superior to others, and he liked to help kids, but often confused his role forgetting he was the student, not the teacher. Also, when Jason did not live up to his own expectations, he would have a meltdown. So Jason's mother enrolled him in after school programs, so that Jason could get more peer interactions in a less formal environment that would mimic the public school arena.

As a result of her involvement, Jason was able to leave our school after two years. Even then, Jason missed his friends and his mother would bring him to the school to visit his classmates when his new school had a half-day in order to help him with the transition process. On one such visit, she saw me and talked about her concern that Jason wasn't getting his counseling time at the public school like he had at my school, and so we discussed ways of dealing with this issue (public school counselors have a heavier caseload and often see kids every other week rather than weekly).

Just as it is possible for a therapist to be hands-on with a student in a constructive way rather than a destructive way, it is possible for a parent to be hands-on in a constructive manner. Unfortunately, when working in a SED, residential, or other psychiatric setting, most parents tend to be hands-on in a destructive manner. Just as you need to set boundaries with your clients, you will need to set boundaries with the parents of your clients. I know a lot of teachers who give out their cell phone numbers to the parents of their students.

Next thing I know the teachers are shaking their head, and telling me how a parent called them on the weekend at 10:00PM at night sounding drunk and incoherent. My response to this is always, "You gave out your phone number? Big mistake." I never give out any personal information. When I call a parent, I leave a message stating what hours I can be reached at the school, and give out the school phone number only. If a parent becomes particularly needy, I tell them, "It sounds like you have a lot of concerns—more than we can discuss over the phone. I suggest you call the IEP coordinator and schedule a parent-teacher conference." Chances are, the parents won't do this. Picking up a telephone and talking all day when you are an unemployed parent is easy; making the effort to take a bus to the school for a meeting is more than most parents are prepared to do. The same is true of teachers. They are being paid to educate the children, not provide emotional support for the families. If parents are calling for things other than a specific issue, suggest parenting classes, or tell them to write down all their concerns and schedule a meeting, so you can give them the attention they deserve face-to-face.

The reality is I am being paid to provide therapy services to the children. Not to the parents. I can suggest therapy to them. But as I already mentioned, most of the parents think they are perfectly normal.

28

Recognizing the Limits of Your Profession

Whether you work with hard-to-reach children in the capacity of a therapist, teacher, child-care worker, social worker, or psychiatric worker, there are many areas of the child's life that you will have no control over. Even with a higher-functioning child living with their biological parent, your control over the child's environment is very limited. As a therapist, I technically work with a child one hour a week. In reality, I may have contact with the child for several minutes a day even on the days when they do not receive counseling. I may observe them in passing, and be aware of whether they are having a good day or bad. I also hear reports from other staff about their behavior if issues come up. But my "direct" contact is limited to an hour a week. That is one hour out of 168 hours. I also only work 42 weeks out of the 52 weeks in the year.

Teachers and aides have contact with a child theoretically 30 hours a week. This means that there are 138 hours that they have no control over. That is also assuming the teacher is in the classroom those entire 30 hours during the week—which is rarely the case. They have meetings to attend, go to lunch, have breaks, and may be outside with a child who is having a meltdown—or are outside talking to a therapist or staff member. Even when a teacher is in the classroom, their attention is divided among all the students. So what takes place during all those hours that a professional does not work with the child?

As mentioned in the previous chapter, the biggest influence over the child's life is the parents—be it a biological parent, legal guardian, or foster parent. Parents have control over their child's behavior and choices for 80% to 70% of the school year, depending on the hours the child attends school and how long they are on the bus. When school is out of session, they have 100% control over their child's behavior. I have had children who were ready to return to public school at the end of the school year who did not attend summer school, meaning that they were off from school and in their parents' custody for three months. When these children come back to school in the fall term, they have regressed to where they were behaviorally a year ago. It takes just three months (or less) of being at home with parents to undo all the behavior modification that was accomplished in approximately one year.

When kids are out of school, there is no way to monitor what treatment they are receiving. Are their parents taking the children to the psychiatrist for medication evaluations? Are they continuing with outside services? Are they continuing to use appropriate consequences and reinforcers at home, or is the child throwing tantrums and getting whatever they want?

Even when school is in session, you have no control over the parents, except to file a report if the behavior appears abusive. If you look hard enough, you can find ways to make it abuse. If the child has it in their IEP that they are supposed to be on medication, but the parents do not give the medication, that is "medical neglect," especially if the child's behavior is so out of control they are a danger to others. If the child is supposed to receive outside counseling but does not, and they are depressed, that is medical neglect because lack of treatment could lead to a suicide attempt, and therefore they could be a danger to themselves.

Unfortunately, DCFS is another area that a professional has no control over. How they respond, and how they handle a case is out of one's hands. The Department of Children and Family Services can keep a case open for years, or close it after two weeks. They can close the case after they make the call and decide there was no abuse. The only thing you have control over is how often and what you report. You may get a social worker assigned to the case who is "involved," and comes to your place of employment to speak to you and get a full case history, or you may get someone who does not care and never contacts you. If you get a worker who actually contacts you, then you find out the case was closed, you can always call the department and ask to speak to a supervisor to complain about the worker. But again, there is only so much control you have.

The probation department is another area you have no control over. I had a student who was on probation who had been in juvenile hall multiple times, physically assaulted a staff member, was blackmailing a student for money, was sexually harassing students and staff alike, but was kept at our school for one year. His probation officer never came to the school and never contacted me or removed the child, even though the principal made multiple phone calls to his P.O. On the other hand, I had a student who was at the school for only two weeks, he had perfect behavior—every time I walked into his class he was the first student with his assignment completed and every answer was correct. He was well mannered, and had no behavior problems. His probation officer came and took him away in handcuffs because they found drug paraphernalia in his home. Mind you, they did not find drugs in his home, just paraphernalia. So apparently a student can assault a staff, blackmail a student for money and be sexually harassing, and he's allowed to stay at home, but smoke a little weed—and oh my goodness—lock him up!

Children are often receiving outside services such as medication through a psychiatrist, wrap-around, TBS, or outside counseling through an agency. Sometimes these workers will contact you to discuss the case and will share information, but often times it is a one-way street. They want information from you, but will not give you information. It depends on the worker assigned to the case, just as with the DCFS. Some kids have a multitude of services (I have one child who sees a different therapist every day of the week), while other students have no services. Often the imbalance of who gets what can be frustrating.

With one of my psychotic children I made repeated attempts to learn what medications he was on since his psychotic symptoms were escalating, but every time his outside worker came to the school she denied having access to the information. I spelled out clearly that it was imperative that this child be on anti-psychotic medication for his safety, and it was like speaking to a wall. She never called me with any information, although every time she came to the school she dutifully took down my name and number.

I had two students who were receiving wrap-around services from the same therapist. She was very helpful, and gave me her card and returned my calls. She spent an hour with me briefing me on both of my cases, giving me background information I would have never received otherwise, especially as one case was Spanish-speaking and I could not contact the guardian. I learned from her that the one case was going to be closed because the student's behavior had stabilized, at which point I told her that in school his behavior was actually escalating. It appeared he was off his ADHD medication, and the honeymoon period of being in a new school was over. Both his teacher and I were having the same problems with him. She said she would do what she could to keep the case open.

Needless to say, I have never once bothered to call a psychiatrist treating any of my students (nothing against psychiatrists—I actually work with one). However, the clients that I have tend to see psychiatrists who spend 5–10 minutes with these kids, bill medi-cal $240, give whatever diagnosis the parents ask for to get the money from the state that they want, and aren't about to return a phone call from a lowly MFT to discuss the case of a child they probably don't remember. As a result, I rarely know what medication they are on (when I call the parents, they somehow can't seem to find the bottles to tell me). And unfortunately, the medication is usually the wrong one. The parents are the ones who explain the behavior problems, but because the parents are not "psychological" experts, and are usually in need of psychiatric treatment themselves, they are not the best resources for information. If psychiatrists really wanted to ensure the clients had the right medications, they would be talking to the school therapists and teachers.

There are psychiatrists who actually spend time with children, specialize in childhood disorders, and will do an in-depth intake before prescribing medication. I had one parent complain that his daughter only saw her psychiatrist for less than ten minutes each month. I recommended one such psychiatrist who was located in a city approximately a 20-minute drive away from the family's residence. The parent's reaction? "That's too far! I can't possibly drive all the way over there!" On the other hand, I once had a parent who drove her son over two hours to see a specialist at UCLA on a monthly basis. Sometimes if you want quality care, you have to make an effort. But as I said, as a professional, you can't control what the parents or the outside professionals are going to do. The sooner you realize and accept that, the sooner you will be able to focus and concentrate your energy on what you can do for the children you work with; otherwise, you will spend a lot of negative energy questioning the system and pulling out your hair in frustration.

In addition to all the outside adults in the child's life, there are their peers. Their friends, classmates, and of course, the gangs. And now, thanks to the invention of the internet, which is how most of my client's appear to spend their time socializing, there are the "friends" they meet over the internet who even my clients do not really know who they are. Students living in gang neighborhoods are under continual pressure to join. There are threats to their family members (both real and empty) if they resist. Students have relationships with friends from former schools, placements, and other assorted settings who have negative influences over them.

I have had clients start using drugs as a result of being in a relationship with a drug-addict. I have had clients miss school because of their involvement with a boyfriend or girlfriend who had to go to court. I have had clients drop out of school when they turned eighteen because their friends convince them they can "get them a job."

One of the best examples I have of how great peer influences are, and also how clueless parents are as to what their children are really doing, is the following. One of my students who typically had perfect attendance missed school due to neck pain, according to the attendance records (I always check the attendance records to know why my kids miss school so that I can confront them on their behaviors when they return). When he showed up the next day, I asked him about it. It turns out his neck pain was from being up on the computer until 1:00AM. Was he working on a homework assignment? No. He and his friend were busy on the computer spamming people.

This student started to miss school at least one day every month, and I suspected it was due to the types of friends he was making on the internet and the activities he was involved in online. It was only when he was reminded that he needed perfect attendance to return to public school that his behaviors returned to normal (or I should say, normal for him).

Peers and gang-members possibly have even more control over my clients than the parents. When a child feels that their parents do not understand or love them, they will turn to their peers for that love. Especially when they are older and their hormones kick in, which is why there are so many pregnant teens at our school. As I mentioned when discussing Jesus, he felt being in a dysfunctional relationship was better than no relationship at all. He turned to his girlfriend for the love he felt he was not getting at home.

An additional element, that can not be seen, or measured, but is often one of the most important keys in determining whether a child succeeds or fails is their level of motivation. Contrary to what one may think, motivation is not a given. Motivation, like any other aspect of a person's personality, runs on a spectrum. Whether motivation is a product of genetics or environment is a moot point for research specialists to debate. The point is that, just as not all children have artistic ability, or musical ability, or scientific inclinations, or athletic prowess—not all kids are motivated to change.

Therefore, I have worked with kids who academically were on a similar level—their cognitive functioning was intact, they were the same age when I started working with them, their family situations were the same in terms of level of function (or dysfunction), their connection to me as a therapist was the same—but one child succeeds, and another child fails. Why? The only difference is that when you ask Billy what he wants, he says over and over, "I want to go to public school." When you ask Bobby, his response is, "I don't care. So what if I stay here?"

In session Billy will ask repeatedly what he needs to do to succeed and reach his goals, while Bobby will play games, talk about who he beat up recently, and his string of girlfriends. On paper, their test scores look the same from a scholastic point of view, but I can guarantee that if administered psychological tests, the kids who fail would show up as lacking in motivation and underachievers. Everyone knows about the over-achievers, the ones who are driven to success, and never stop—they get their degree, then go for a second degree, and when they get a job they get promotion after promotion. No one questions where they get their drive from because they are successful. But when someone leaves school we label them lazy. However, through my private practice in working with adults, I have learned that there are enough adults out there who are content doing what they do, and it is not so much that they are lazy, as they are unmotivated for intellectual growth. When working with this population you have to understand that motivation—like artistic ability, athletic ability, and musical talent—cannot be taught. If a child was tone deaf, you would not force him to play a musical instrument. Some kids are happy with the idea of *not* getting their GED, not going to college, and working a menial job for the rest of their life. If you plan to work in a school, you need to understand that it may not have to do with the student being lazy; they may lack the motivation gene.

And last but not least, there is the environment that you work in. While it would seem that a professional worker would have some control over their own work environment—after all, the environment is geared in the best interest of the child, that is not always the case. Schools are controlled by governments, who would rather build jails than spend money on education or early intervention. How the money is spent within the school is controlled by an administrative department, who often has no understanding of what is and is not important in long-term versus short-term behavioral success.

The group home where I worked for two years terminated all the therapists because they wanted to find a cheaper means of getting therapeutic services for the children. Yet, as I mentioned in the chapter on attachment issues, these kids had gone through multiple therapists and I was the only one to last there that long. That, apparently, meant nothing to the group home. Helping the kids learn to form healthy attachments to role models was not as important as saving money.

The only thing you have control over is choosing where you work, and how involved you want to get in the politics of the environment. When I worked in Watts, the teachers had no respect for therapists. We were seen as incidental staff who "played" with the children. I think in my two years there I was sought out only once to consult with a teacher regarding a student. None of the teachers knew my name or the names of my co-workers, some of whom had worked there for over five years. I was referred to as "Hey you, counselor." I was often prevented from filing DCFS reports, which is against the law.

At my current school, the teachers, counselors, and IEP coordinators have a team approach. The staff—both the educators and administrators—always communicate with me regarding my students. Do I agree with how money at my school is spent? No. Do I always agree that the classes my students are placed in are the best ones to meet their academic needs? No. But I have never heard of a school that is perfect, and given the benefits of this environment over others I have worked at, this particular school works for me.

Basically, the only thing you have control over, regardless of what profession you work in, is the time you spend with your child. You can always make recommendations to the appropriate authorities or people in power, but after that, the ball—as they say—is out of your court.

29

The Ones That Got Away

While I do not keep track of the actual statistics, I would guess that for every child I work with for multiple years until their case is closed—either because they return to public school, graduate, or move—there is another child who disappears off my case load for less fortunate reasons. They could quite literally disappear and you have no idea what happened to them. They could go to Juvenile Hall, jail (if 18 or they have been to the Halls too many times and the court determines jail is more appropriate), they could be placed in a more restrictive environment such as an out-of-state camp or locked facility, or they could be removed from the home and transferred to a school closer to their new facility.

Which brings me to the subject of termination. In normal private practice, when working with a "healthy" individual, a therapist and the client jointly agree that it is time to end therapy, and will have time to terminate and discuss with the client how it will feel to no longer see the therapist each week. When working in a school, this opportunity rarely happens.

If a student is returning to public school or graduating, and you know in advance when they are leaving, you can have a transition process in therapy to discuss how they will cope with their new school. You can also discuss how they will feel about not seeing you every day (on campus). For kids who have difficulty expressing feelings, this may be the most challenging session of their life.

Unfortunately, most of the time there is no termination session. You will arrive at school and see on the attendance sheet "hospitalized, no further pick-up" or "moved out of state." Or you will find out that the police came the day before and carted your student off in handcuffs to a facility in Colorado or Arizona. There is no chance to say goodbye, let alone have a termination session. There is no warning that they are leaving. However, it is not the student who feels upset about not being able to say goodbye, because they're in handcuffs and pissed off. It is you, the clinician or teacher, who are upset because you had a relationship with someone, and now that someone is gone.

This is one of the reasons why setting up good boundaries is important if you are going to work with these kids. They can disappear quickly, and if you become too attached, it can take a piece of your soul away every time you lose one. The good news is, that after working with these kids long enough, you can start to recognize which ones won't last in the environment you work in, and will eventually "disappear." Occasionally, however, some will take you by surprise.

One little boy I worked with, Mark, was shy and reserved. He rarely spoke in session, but was always polite and cooperative. After working with him a year, I saw great improvement in him. He smiled and laughed more, and was more playful with his peers on the playground. Suddenly, his attendance dropped for no apparent reason. His mother started making complaints about the school. She made allegations of physical abuse to her son, none of which were substantiated.

Because Mark was not attending school, truancy reports and child abuse reports were filed against the mother. It turned out that she had removed her son from the previous school for the same reason, and had sued the school. Well, sure enough, she sued our school as well, and Mark disappeared. We never knew where he went or what happened to him. Technically, when a child is enrolled in a new school, they request the records from the previous school. No doubt, Mark's mother took him back to Mexico when she was unable to get the money she was looking for through a lawsuit.

Clark, was a typical wanna-be gang-banger. Unlike the actual gang-bangers, Clark boasted all the time about being in gang. He tagged gang signs all over his desk, his school work, and his clothing. Clark had issues with his family, but would not discuss them in counseling. He only wanted to attend counseling if it got him out of a subject he hated. He had the swaggering, "I don't give a fuck" attitude of a kid heading for the Halls. Clark would miss school for weeks on end, then show up, each time with a different story about what was going on in his life. It was impossible to sort fact from fiction. At Clark's IEP, I recommended an AB3632—residential placement.

Clark was unable to attend school on his own, and his behavior appeared out of control. The placement never took place. His attendance continued to be erratic, and he would disappear for two to three months at a time.

This behavior went on for two years. Eventually, Clark was sent away to a facility. Unfortunately, the older a child is once they enter residential placement, the less success there is. If he had been placed when he was younger, his prognosis would have been better. By the time he was placed, he had actually joined in a gang and was using drugs, and it was unlikely that the residential staff could undo the damage that had been done by these influences.

Another student who was sent away to a facility was Marshall. Marshall was overweight, in a gang, possibly doing drugs, but the biggest problem with Marshall was his size. Since he was grossly obese, he was able to use his size to intimidate people, including his own family. Marshall participated in counseling about half the time. It depended on his mood. Sometimes he would discuss things openly, and sometimes he would not. I knew Marshall loved his mother, but he had issues with his father. Marshall was one of the students in my office who was always using my stencils to create pictures for the staff at the school or his myriad of girlfriends—so I knew that on a deeper level Marshall cared.

However, at home Marshall's behavior was out of control and his mother requested he be put in residential placement. An IEP for an AB3632 was held, and during the meeting, when Marshall's mother discussed all the problems she had with Marshall—his staying out all hours at night, stealing her car, being physically abusive towards her and her husband, her fears of safety for the smaller children in the home—the school psychologist told her that the next time Marshall acted up to call the police and they would take him away. She was informed that since Marshall was "special education" the police would respond to the call more urgently than a regular call and pick him up right away. Meanwhile, while hearing all of this, Marshall started to cry and ran out of the IEP meeting.

The following week in counseling, I took Marshall for a walk around the block. I told him about how I worked in a group home, and what his life would be like if he lived in a residential or group home facility. I asked him if that was what he wanted. Marshall replied in the negative. I told him, quite simply, "Then get your shit together and stop acting like a bully at home, or that's where you're going to end up." I pointed out to him all his great qualities, his sensitivity when he did art work for the staff, and that I knew he cared, so he needed to work harder in counseling to deal with his anger issues or his mother wasn't going to put up with his shit.

Marshall had truancy issues, and would disappear for weeks or months, then attend school for months with perfect attendance. Needless to say, Marshall's mother, like many parents, could not bring herself to put her baby (who weighed more than her and was making their home a living hell) in residential placement. Things got bad again, and she requested Marshall be placed, but by that time his AB3632 had run out—when a case is opened, it is open for so long before the case gets closed and the process has to be started all over again.

Marshall ended up eventually in Juvenile Hall, and came back to school under house arrest. Apparently that was not enough to convince him to go on the straight and narrow, and after getting off of house arrest, less than one year later, the police showed up at our school and took Marshall off to a locked facility out of state. As with the case of Clark, if Marshall had been placed in a residential facility at an earlier age, he probably could have turned things around more quickly. Marshall was, at his core, a good kid. By the time Marshall left, he was seventeen, and his mother was acting out of desperation before her son ended up in jail, not Juvenile Hall.

One client who lasted less than a month at the school was Erika, a bipolar teen who was pregnant. Because she was pregnant, she was not on her medications. Medications used for bipolar disorder and schizophrenia can cause abnormalities to developing fetuses. In addition, because she was pregnant, the staff was not allowed to physically restrain her if her behavior became out of control. Counseling sessions were unproductive due to her mood swings, and around school, she was seen continually being physically inappropriate with boys. Within a month, she was transferred to a locked facility since she was a danger to herself and others.

A client who managed to stay at our school much longer than he should have given his rap sheet, was a young boy, Mario, who was on probation. Mario was around twelve, and had been in and out of

Juvenile Halls numerous times. His father was in prison, so apparently Mario felt this was the "manly" thing to do. Mario physically assaulted a staff member with a bunch of his gang buddies at school. I do not mean "gang" in the sense of gang-members, but he had attracted others, like himself who were physically bullying, and liked to prey on students who were perceived as weak. He and two others would sexually harass and physically assault students and staff alike.

The school separated the three students, putting each one in a different class, and eventually the other two boys were sent away, but Mario remained. The school continually advised Mario's probation officer that Mario was in violation of his probation, as did I, but he continued to stay in the school rather than get placed in a locked facility. After I spoke to Mario's probation officer about his behavior, Mario requested a change in counselors, which was fine with me. His behavior towards me was sexually offensive, and he had never showed the least interest in working in session. He came only when it suited him, and never showed any remorse for his actions (this is where the diagnosis of conduct disorder would come in). After approximately one year at the school, Mario disappeared. I have no idea where he went, since I was no longer his therapist, but I suspect Mario will end up in jail before he turns eighteen.

Finally there was Melissa. Melissa had multiple outside services, including wrap-around, outside counseling, mental health services at the school, and me. Melissa did not click with me. Melissa was deeply disturbed, depressed, probably borderline, and clearly suicidal. However, she seemed able to manipulate all the other professionals on her team. There was a wrap-around meeting held, and all the team members were talking about all the progress Melissa was making, and I thought, "Am I missing something? Are these people blind?" However, because I had only been working with her for a month, while the other people had known her for over a year, I kept my mouth shut.

Melissa was almost eighteen, and had all these plans on what she was going to do when she turned eighteen—none of which were very good choices. In session I would confront her on these choices constantly, and she would shut down, and say I didn't understand.

The problem was, I understood Melissa better than anyone else, and that scared her. Melissa eventually said she felt she didn't need two therapists at the school, and stopped seeing me. She continued to see the other therapist who she was able to manipulate more easily. Melissa eventually ended up cutting both her arms up, and was hospitalized. While she returned to school briefly, once she turned eighteen I noticed she disappeared from the campus. Whether this was because she dropped out or was locked up, I don't know because I was no longer her therapist. Unfortunately, this is often the case. The seriously disturbed students who know that you understand them to the core, are afraid by such a deep understanding and run from it.

Sometimes a client that I worked with for years will disappear, and I will hear from the student grapevine about where the student is or what they are doing. I take this gossip with a grain of salt. Sometimes I hear it from multiple sources, but again, that could be word-of-mouth from student to student. Occasionally, I will actually get a phone call from a student after they leave.

For the most part, once a student leaves, I have no idea what happens to them. They can pop back up on my caseload two months or two years later. I could hear rumors that they died on the streets. Occasionally, I'll get notified by the LAUSD computer system that the student is at another school. I try not to think about the ones I did not help, anymore than I think about the ones I did help. I tell the ones who leave me to go to public school, "I didn't help you do this. *You* did this. I gave you the tools, but *you* made this happen." I do not take responsibility for my kids' successes, and I can't take responsibility for their failures. If I did that, I could not go to work each day.

30

Preventing Burnout

Whenever a new teacher starts at our school and I introduce myself as one of the therapists, I inevitably get asked, "You're licensed? Do you have any advice for me?" The teachers at our school are usually working on their credentials and taking their exams. Many of the classroom aides are getting their degree. I always tell a new teacher the following, "Don't take your work home with you emotionally, and only expect to help one child in your class during the entire school year." When I state the second part, the teachers always give me this concerned look, as though I am crazy. I'll reiterate it, saying, "Really, these children are very disturbed, and if you think you can make a difference in all twelve of their lives, you are going to quit after one year. Have very low expectations and you won't be disappointed. You can only help one child in the school year."

I remember when Warren, a nice young teacher who was assigned to a high-functioning autistic classroom met me. He clearly thought I was nuts. He tried so hard with his kids. He brought in his guitar, singing songs to them, trying to get them educated about music, art, and areas that generally weren't taught in our school. One day during the summer break I came into the school to pick up my paycheck, and I passed Warren's classroom. He was packing boxes. I walked in, commenting, "You're leaving?" We then went on to have a discussion about what I had said to him when he first started working at the school. He had remembered my advice, but had not taken it to heart.

He continued packing his supplies. "I should have listened to you. I let the kids get under my skin. I can't work in this environment. I found another school where the kids are less disturbed."

I shook my head. "I told you not to take your kids home with you." The sad thing was, Warren was a good teacher, and he often had in-depth conversations with me about the students I had in his class.

He really was trying to learn and understand about the psychological side of their disorders. Unfortunately, he had set no boundaries with the parents, had been abused verbally by many parents, and was overwhelmed by trying to do too much for too many.

I've had this conversation with many teachers—even those who have worked at the school for years. I remind them that they are up against the influences of the parents, the gangs, the system, and they can only do so much. But for some reason, teachers want to save every child in their class. When I worked full time at the school and had over thirty students on my caseload, I held to the same rule—one child per year. If I could help just one child per year, I was doing my job. That was the best I could hope for given that the odds were stacked against me. I now work only part-time, and I still keep the same rule.

The children at SED schools, group homes, and residential facilities often leave abruptly. I know many therapists or teachers who spend many hours of their time following up with their client to find out where their client went and what happened to them. Personally, once the student is off my caseload, I do not think about them again. It is not because I do not care; it is because if I spend too much time caring for the ones I have no control over—the kids who disappear into the system, then I do not have the emotional strength available to give my attention to those clients who remain within my care. Often you hear rumors of children being homeless, on the streets, in locked facilities—schools are a grist mill for rumours. If I wasted my time chasing down every rumour that came across my desk I would be an emotional basket case in need of therapeutic intervention myself.

I remember one year shortly after I had been licensed and it was near summer break. One of the teachers I was friendly with, Jessica, was sitting while the buses unloaded, waiting for her students. She made some off-color comment—I do not remember her exact words, but the comment surprised me. Jessica was an excellent teacher. She was able to keep her class in control, and her kids respected her. I continued walking, but then stopped, and went back to her. I asked her if everything was all right, and she made another remark that was negative—almost hostile. Not hostile to me, but hostile about the students. I touched her shoulder, leaned over and whispered, "It sounds like someone is suffering from burnout."

She laughed bitterly and said, "You may be right."

Jessica did not return to school the following September. She took a year sabbatical and returned a year later. She told me my comment had made her think, and she realized that she *was* burned out. Part of it was due to administrative problems beyond her control, but when you are inside your own skin, sometimes you can't see what others see. Normally Jessica was all smiles and motivated her kids through a positive attitude. Seeing her sitting there slumped in defeat concerned me about her mental well-being.

When I tell people what I do for a living, I'm often told I must be a saint and have a lot of patience. I am neither a saint, and I definitely do not have a lot of patience. Just ask anyone who has been stuck in a car with me during traffic on a Los Angeles Freeway. So how have I survived this work for nearly a decade without quitting?

As I said—I do not take my kids home with me. When I leave the office, I do not think about them at home. I have friends from work, and we have a "no work" rule when we get together. If either of us starts to talk about work, we cut the other one off, and say, "no work." We keep boundaries so that friendship and fun time is that—fun time. I remember one time before a summer break, when I had a cruise planned, one of my kids asked me, "Will you think about me when you're on your vacation?" I told him, quite honestly, that it was nothing personal, but that if I was thinking about my students on my vacation then *I* needed to see a therapist.

Obviously, there can be exceptions. When my client died after I filed multiple abuse reports, I called several therapist friends of mine to vent. I allowed myself a week to wallow in anger and frustration, and then I moved on. Why just a week? Because any longer, and I would not be able to do my job appropriately at the school. I had to help out the kids who had been friends with the boy who died, and help them through their grief.

Which brings up having a good support system. I have friends who work at my school, and I also have friends who are therapists. It helps to have friends in the profession you work in, because when you want to vent about something that happened at work, or you need to confer on a difficult case, they understand and speak the language. I also have a lot of friends who do not work in my field, and they are great for other areas of support, but in a situation like a client dying, or being removed from the school after years of being on my case load and I do not get to terminate—there is nothing like having a colleague who has been in a similar situation to talk to.

Additionally, while there are many things at my school that I have no control over, one thing I can control is how I stagger the clients on my caseload. Obviously, if I saw six depressed clients in one day back-to-back, that would be emotionally draining, having one child after another reaching for a tissue while they cry throughout their hour. I tend to stagger my clients so that I see a high-functioning, mid-functioning, and low-functioning client. I do not see two clients back-to-back with the same diagnosis, because that can be too mentally draining. Again, there are obvious exceptions, such as if a client is in crisis and calls requesting to see me, but that rarely happens. This way, my day is balanced so that I am not seeing the same type of client all day. For another clinician, this may not work—they may not want to switch hats, and prefer to see all similar clients back-to-back. Everyone has to find what style works for them.

In addition to finding what style works for you, is finding what environment works for you. Every setting has its strengths and weaknesses. As I mentioned, the school I worked at in Watts had teachers who did not value the clinicians, and taught racial intolerance—but they had a time-out room for kids who misbehaved which helped to maintain control in the classroom. My current school has teachers who value clinicians, but no time-out room which results in the misbehavior being rewarded by the student having an audience to their tantrum.

A colleague of mine worked at a school where they were expected to have their students sign in and out of their sessions. The insinuation was that the clinicians weren't trusted to be doing their job—the problem? Most of the kids were autistic and could not tell time. She was also required to provide her own computer and printer for her office for her paperwork! Another friend of mine developed carpal tunnel syndrome after six months of her internship due to the amount of paperwork she was required to do for the Department of Mental Health. A fellow student from my graduate school informed me that at his internship he was expected to falsify his records which is an illegal offense and can result in a loss of license. At two of my internships, one of which I did not receive any pay for, I was required to be on call 24/7. This resulted in receiving phone calls in the middle of the night which would cause my heart to race. It would be hours after the emergency phone call before I could return to sleep because I was a "newbie" and had little experience handling emergencies.

Every professional must learn through their early experience what environmental work conditions they are willing to tolerate and what they won't. While my current caseload consists of more severely disabled clients, I do not have to worry about my phone ringing at 3:00AM. Once I leave the school, my work is done.

I can't reiterate enough the importance of recognizing countertransference issues—as a clinician, teacher, child-care worker—whatever your profession. If you find yourself thinking about a client too much—ruminating on that student or client excessively, feeling hostile or emotionally attached significantly more than you are with the other students in your class—seek professional help. It may only take a few months, but it is important to understand where this stems from, otherwise, the same situation will occur over and over again, with different students passing through your class. Everyone has students or a particular population who pushes their buttons. Anyone who says they do not is in denial. I am appalled by the number of therapists I work with who say, "Oh, I can work with anybody!" Usually, what they are saying is that they are so desperate for money, they will work with anyone, even if the sessions are unproductive because they are unable to establish a therapeutic relationship.

Many therapists, because they do not get paid benefits, will take any client on their case load, regardless of whether or not it is a client they work well with. Other therapists have a martyr complex, and think they are a better person for working with all these "difficult" or "unwanted" cases. These therapists are usually the ones without good boundaries, who not only are working with their own students, but are forming relationships with other therapists' clients. The problem with both of these situations, is that the therapist then starts to resent that they are seeing these "difficult" cases, when they should have turned them away—or recommended that counseling be discontinued (as I did in the case of Mrs. X), or they are so busy "counseling" students who are not their own that they feel overwhelmed and put upon.

Without appropriate boundaries, and the ability to recognize what your issues are and how they are impacting your professional work performance, burnout is almost inevitable.

As a therapist, I have to take care of myself first and foremost, before I can be there for others emotionally (and my work takes a lot on the emotional level). I make sure that I eat healthy and regularly.

I actually take my lunches in to work in order to assure that I get a balanced meal and am not eating food with too little protein or too many additives. I keep water bottles in my desk so that I can stay hydrated at work (especially during the summer when outdoor activities often take place and the temperature reaches over 100 degrees).

I make sure that I exercise regularly. At one point I took a third job which was only five hours a week, but required me to be on the freeway at 7:00AM. While I loved the job, and it paid great money, it interfered with my exercise time. After three months of not exercising, staff at my school started making comments about me. I wasn't smiling often. I looked tired. I looked grumpy. Was something wrong? I realized that as much as I loved the job, it was not worth the impact on my health. I gave notice, and returned to my morning exercise routine. Within a month, I had my energy back, and felt happier and less irritable.

During my school breaks, I use the time to make all my doctor's appointments. At the school, I use a homeopathic spray on my desk and doorknobs made from a solution of water and rosemary oil as a disinfectant since many of the kids do not wash their hands and have hygiene problems. Additionally, many parents send their kids to school even when they are sick and running a temperature—especially the foster parents. So I try to maintain my health as much as I can, not only through diet and exercise, but by getting my regular check-ups when school is out and I have more time on my hands.

I often joke with people that I work hard and I vacation hard. I try to get away whenever school is out of session—usually to nature. I try to find ways to do things that cost little money, although I have taken expensive vacations as well. But when I go on vacation, the most important thing for me is that no one can reach me by phone, and there is no computer around. I like to take vacations where I am physically active since being a therapist I tend to sit a lot. I like to go hiking, or walking, try new experiences. Before starting my job at the school in Watts I spent over three weeks in France by myself, driving around the countryside, going in a hot-air balloon, taking a boat down a river in the Loire valley.

Being self-employed and working two jobs (one of which requires being on the computer), taking time off and being at home is not a vacation. It is too easy to jump on the computer and work on my taxes, or do some sort of work. While I have a great ability to set boundaries at work, I do a terrible job of setting boundaries at home. The only way I can *not* work, is to be away from my home. Since my work requires talking to people continuously, vacationing in nature is the best solution. It's not that I do not like being around people, but I need the solitude of nature to quiet my mind after continuous months of listening to people's problems and working in an environment where there is constant screaming and yelling.

In addition to taking a vacation every year, I try to keep a balanced life. When not working, I pursue other interests and pleasures. Whether you are interested in sports, music, gardening, or knitting, it is important to have areas of interest that you can participate in with people outside of your field. Having something to talk about besides teaching, counseling—or whatever field you work in, is important for your mental and emotional health when you deal with this type of population. Again, you cannot preach to your clients about having a healthy lifestyle when you yourself are sitting at home, mentally suffering because your career is dragging you down.

So how do I measure success in a job where I often do not know what happens to my children? Success with this population isn't always big and news worthy. Sometimes there are big moments, like when Andy returned to public school after being in an SED setting for ten years. Other times, it can be the smaller moments, like when a student of mine who rarely spoke of anything of importance, spent his entire session discussing the problems he was having with his new teacher. Or when a student calls you and asks to speak to you because they are in crisis and you avert the crisis with an intervention. Sometimes it is as small as getting a student to agree to come to counseling who has been refusing to attend sessions for weeks on end. Or convincing a student to attend summer school rather than sit at home for three months watching television.

If you set your goals small, the rewards will be greater and the disappointments less. Then there are the times when a child thanks you for their session when you return them to class. Or they write on your white erase board, "Jillian is the best therapist ever." If you appreciate the small moments, instead of focusing on the big picture, which you have no control over, the job can be incredibly rewarding. Having a small child put their hand trustingly in yours as you lead them down the hall to your office can be the best feeling in the world.

References and Resources

25th Annual Report to Congress on the Implementation of the Individuals with Disabilities Act, Volume 1 (2003). Prepared by Westat for the Office of Special Education and Rehabilitative Services United States Department of Education.

Adoption and Foster Care Analysis and Reporting System (AFCARS), data submitted for the fiscal year 2006 (10/1/05 – 9/30/06). United States Department of Health and Human Services.

Bowlby, J. (1988). *A Secure Base – Parent-Child Attachment and Healthy Human Development*. Harper Collins (Basic Books).

Campbell, R.J. (1981). *Psychiatric Dictionary, Fifth Edition*. Oxford University Press. New York.

Capacchione, L. (1989). *The Creative Journal: The Art of Finding Yourself*. NewCastle Publishing Company. North Hollywood, California.

Corey, G. (1996). *Theory and Practice of Counseling and Psychotherapy, Fifth Edition*. Brooks/Cole Publishing Company.

Department of Health and Human Services, Centers for Disease Control and Prevention. (2007). *Sexually Transmitted Disease Surveillance*.

Department of Health and Human Services, Centers for Disease Control and Prevention. (2009). *Preventing Teen Pregnancy*.

Department of Health and Human Services, Centers for Disease Control and Prevention. (2007). *The Obesity Epidemic and United States Students*.

Department of Health and Human Services, Centers for Disease Control and Prevention. (2006). *Childhood Obesity*.

First, M. (ed). (1994). *Diagnostic and Statistical Manual of Mental Disorders, Fourth Edition*. Washington, D.C.: American Psychiatric Association.

Fleming, J. (2007). *The Food-Mood Connection.* School for champions.com.

Gorman, J. (1997). *The Essential Guide to Psychiatric Drugs.* New York: St. Martin's Griffin.

Grandin, T. (2006). *Thinking in Pictures: My Life with Autism (Revised Edition).* Vintage Publishers.

Greenfeld, D. (1994). *The Psychotic Patient: Medication and Psychotherapy.* New Jersey: Jason Aronson, Inc.

Greenspan, S., and Wieder, S. (1998). *The Child with Special Needs: Encouraging Intellectual and Emotional Growth.* Massachusetts: Perseus Books.

Greist, J., and Jefferson, J. (1998). *Obsessive-Compulsive Disorder Casebook* Revised Edition. Washington, DC: American Psychiatric Press, Inc.

Haag, P. (Ed.) (2000). *Voices of a Generation: Teenage Girls Report About Their Lives Today.* Marlow & Company. New York.

Hayden, T. (1981). *One Child.* Avon.

Hunter, A. (1999). *The Sanity Manual: The Therapeutic Uses of Writing.* Kroshka Books. Commack, New York.

James, B. (1989). *Treating Traumatized Children: New Insights and Creative Interventions.* The Free Press.

Kaufman, B. (1979). *Son-Rise.* Warner.

Landsman, J. (Ed.) (1994). *From Darkness to Light: Teens Write About How They Triumphed Over Trouble.* Fairview Press. Minneapolis.

Lumeg, J., Gannon, K., Cabral, H., Frank, D., and Zuckerman, B. (2003). *Childhood Obesity and Behavior Problems Linked.* University of Michigan Health System Press Release.

National Institute of Mental Health, Mental Health Medications. Available online at http://www.nihm.nih.gov/health/publications/mental-health-medications/

Preston, J., O'Neal, J., and Talaja, M. (1998). *Consumer's Guide to Psychiatric Drugs.* New Harbinger Publications.

Puzzancher, C. and Kang, W. (2008). Juvenile Court Statistics Databook. Online available at http://ojjdp.rcjrs/gov/ojstatbb/jcsdb/

Riley, S. (1999). *Contemporary Art Therapy with Adolescents*. London: Jessica Kingsley Publishers.

Schwartz, J., and Beyette, B. (1996). *Brain Lock: Free Yourself from Obsessive-Compulsive Behavior*. New York: ReganBooks.

Shandler, S. (1999). *Ophelia Speaks: Adolescent Girls Write About Their Search for Self*. HarperPerennial.

Snyder, H. (November 2008). *Juvenile Arrests 2006*. Juvenile Justice Bulletin. The Office of Juvenile Justice and Delinquency Prevention.

Wong, M. (1983). *Nun: A Memoir*. San Diego: Harcourt Brace Jovanovich.

Glossary

AB3632 – A referral to the Department of Mental Health to have a child evaluated for removal from the home and placed in a residential facility after other methods (school counseling, family counseling) have failed to correct a child's behavior.

ADHD – Attention Deficit/Hyperactivity Disorder – a psychiatric disorder resulting in an inability to focus on school work, and an excessive level of activity. Can be diagnosed as ADD (Attention Deficit Disorder) if the criteria for hyperactivity is not met.

Alogia – Lack of speech due to a cognitive disorder or confusion.

Autism – A psychiatric disorder diagnosed in early childhood manifested by a delay in speech; lack of social development; and repetitive, obsessive behaviors.

Asperger's Disorder – A psychiatric disorder diagnosed in early childhood manifested by a lack of social development; and repetitive, obsessive behaviors, but there is no delay in speech and no delay in cognitive development.

AWOL – When a student has left a classroom, or school grounds, without permission and without adult escort.

Bipolar Disorder – a psychiatric disorder resulting from a chemical imbalance that causes mood swings; consisting of both manic and depressive phases, which is treated through medication and therapy.

Borderline Personality Disorder – A personality disorder, traits, or features that result in a person having difficulty forming and sustaining relationships due to a fear of abandonment. Borderlines are characterized by their excessive neediness, dramatic acts for attention when abandonment is perceived (e.g. a suicide attempt when therapist is about to go on vacation), and a tendency to idolize or devalue the people they are in relationships with.

Conduct Disorder – a psychiatric disorder where the predominant feature is the violation of the basic rights of others (including destruction of property and harm to others) with a total disregard for the consequences of these actions.

DCFS – Department of Children and Family Services – Government agency who investigates allegations of child abuse, and oversees cases where a child has been removed from the home and has been assigned a social worker.

Delusion – A false belief that is rigidly adhered to despite the fact that it is contradicted by reality, typically found in psychotic patients such as schizophrenics.

Echolalia – A speech pattern where a person repeats what the other person says, or repeats something they have been exposed to (movies or television for example) which does not fit the context of the conversation and is a poor form of communicating feelings, needs, and wants.

Encopresis – The repeated passage of feces into inappropriate places (e.g., clothing or floor) whether involuntary or intentional; and the chronological age is at least 4 years or the equivalent developmental level.

Functional Analysis Assessment/Behavior Intervention Plan (FAA/BIP) – A report done when a student is at risk of suspension or has been removed from current setting for more than ten days. It examines what interventions have been tried, their effect, the motivations to child's behavior, and whether current behavior supports are working. Also known as Functional Behavioral Assessment (FBA).

IEP – Individualized Education Plan – A yearly meeting for a student with a disability required by law where treatment goals for the student are developed, and each year the student's progress towards the goals are discussed and new goals are set forth.

Incident Report – When a child commits a serious offense, either physical, sexual, or threatening; a report of his behavior goes into his file and home to his parent or legal guardian.

LOP – Loss of Privilege – when a student has earned sufficient marks for misbehavior, or been inappropriate in a severe enough manner, such as physical aggression, to lose privileges for a period of time and earned a consequence (usually an additional academic assignment).

Obsessive-Compulsive Disorder – A psychiatric disorder by either obsessions (recurrent and persistent thoughts, impulses or images) or compulsions (repetitive behaviors which a person feels compelled to carry out). The person recognizes that these behaviors are excessive or unreasonable (in children this may be absent).

Oppositional Defiant Disorder – A psychiatric disorder characterized by a child's disregard of parental/school rules, but does not cross over into the violation of basic human rights.

Schizophrenia – A psychiatric disorder characterized by delusions, auditory hallucinations, and/or visual hallucinations which is treated through medication and some type of supportive psychotherapy services (group, family, or individual).

Social Phobia – A psychiatric disorder predominantly characterized by a persistent fear, that is excessive and unreasonable, in social settings that are familiar, and in children this fear is expressed by crying, tantrums, freezing or shrinking away.

Tangential – A speech pattern whereby a person tends to ramble on from topic to topic randomly, without waiting for a response by another person. Often, the topics discussed have no cohesive relationship to one another, and therefore the flow of dialogue becomes disconnected and convoluted.

Tourette's Disorder – A childhood tic disorder with multiple motor and one or more vocal tics have been present and occur many times during the day, usually in bouts.

Trichotillomania – A psychiatric disorder characterized by the pulling out of one's own hair resulting in noticeable hair loss.